31 PRACTICAL ULTRALIGHT AIRCRAFT YOU CAN BUILD

BY DON DWIGGINS

FIRST EDITION

SECOND PRINTING

Printed in the United States of America

Library of Congress Cataloging in Publication Data

Dwiggins, Don.
31 practical ultralight aircraft you can build.

Includes index.
1. Airplanes. Home-built. I. Title.
TL671.2.D893 629.133'343 80-14764
ISBN 0-8306-9937-6
ISBN 0-8306-2294-2 (pbk.)

Cover photo courtesy of Larry Collier.

Contents

Introduction

The decade of the 1980s has come at a time when the pendulum of technology's time clock has already begun a giant swing away from the search for bigness toward a beat more compatible with the economics of ecology. In the 1970s, we all became more aware of the limitations of the natural resources of our small planet.

The sudden emergence of ultralight aircraft in the burgeoning homebuilt movement is a new direction for experimental design. Hundreds of thousands of young men and women in the United States and abroad are joining the action. By one Federal Aviation Administration official's estimate, today there are more than a quarter-million people actively involved in the process of building and flying the most rudimentary of ultralights—hang gliders, both powered and unpowered.

There are many reasons for the emergence of the "minimum airplane" at this time—socio-economic, technological and sheer fun. For the first time in the long history of aviation, we are approaching the elusive goal of flying with the freedom of the birds, the way Leonardo da Vinci envisioned centuries ago.

Wide-bodied jets have their place in the sky, for all their technological problems, and so do civilian and military supersonic aircraft. Multi-seat private aircraft, communter planes and other types of general aviation transportation machines will continue their popularity, of course. But now as never before, the trend is toward small, economical, inexpensive aircraft that don't cost an arm and a leg. Instead of paying $25,000 for a small commercial two-seater air trainer, many pilots are flying lovely homebuilt craft that cost only one-tenth as much or less. And they are having as much, if not more, fun.

We are sure to see new, exciting developments in the immediate years ahead that seem certain to revolutionize the world of flight. Exotic new materials are coming into widespread use in the homebuilt movement. New aerodynamic designs are pointing the way toward revolutionary concepts that seem certain to help the ultralight explosion fulfill its promised destiny—the small, safe, fun aircraft for everybody.

The 1960s closed with great promise for manned conquest of space following Apollo 11's 1969 lunar landing. Space conquest slowed during the 1970s for some of the reasons that aviation technology appears destined to make a 180-degree turn in the 1980s. Big is no longer beautiful.

Don Dwiggins

Chapter 1

A New Sport is Born

Ultralight aircraft are not new. The first powered flying machines were of necessity ultra-light, compared with today's popular general aviation aircraft. There were several reasons for this, the prime factor being non-availability of powerplants with adequate power-to-weight ratios. The first airplanes were built of lightweight bamboo and silk because sheet aluminum was simply not available. The early designers and experimenters used birds as their models. They knew that heavy birds like the ostrich and emu were virtually ground-bound by their own weight.

The rebirth of ultralight powered aircraft had its roots in the reappearance and acceptance of unpowered hang gliders in the 1960s. It was almost a replay of a scene of the early 1900s when the Wright brothers added a crude powerplant to a Chanute-type biplane glider at Kill Devil Hill, to launch the era of heavier-than-air powered flight.

Modern Gliders

The modern hang glider movement got under way at Langley Research Center when a National Aeronautics and Space Administration engineer named Francis M. Rogallo developed a flexible-wing para-glider for military applications. In Australia, the device was quickly adopted for sport flying. Bill Bennett first used the Rogallo wing glider for aerial jaunts behind a speedboat, while in tow on water skies. He soon introduced the sport to America.

About the same time, another revival of hang gliding occurred in Southern California. A school teacher, Jack Lambie, helped his sixth-graders build a Chanute-type aircraft with 28-foot wings from $24.95 worth of scrap wood, wire and plastic. The kids called her *Hang Loose* (Fig. 1-1) and were delighted to watch Lambie make a number of short flights down a hillside near San Juan Capistrano Mission.

Other biplane hang gliders quickly appeared in Western skies, notably one graceful, swept-wing, tailless craft named for Icarus—the legendary Greek youth who flew too close to the sun and melted his wax wings. *Icarus* was the creation of a 16-year-old son of an aerospace engineer and astronomer, Taras Kiceniuk, Jr., who became the envy of the Rogallo flyers by making extended flights along coastal slopes where the ocean breezes provided steady orographic lift.

A refinement of Kiceniuk's design, called *Icarus II*, soon became known on hang gliding slopes the world over. But it shared one drawback with all other hang gliders—you needed a hilltop launch site to get airborne, with gravity doing the work. John Moody, of Milwaukee, WI, decided to do something about that. In the mid-1970s, Moody was an experienced Icarus II pilot with a brilliant idea. Why not add an engine and dispense with re requirement of a downhill ground-skimming launch?

A Powered Glider

Moody installed a McCulloch 101 two-cycle engine in his Icarus II, and soon was foot-launching the machine from flat Midwestern fields with the greatest of ease. On one flight, using a backup plastic-bottle fuel tank, Moody set an unofficial world altitude record for powered hang gliders of 8700 feet above mean sea level after launching from a field at an elevation of 690 feet MSL. The record flight was accomplished without the aid of thermal, wave or ridge lift.

A major goal for Moody was to make hang gliding in ultralight machines safer than ever. With a power package installed in an ultralight machine, a beginning pilot could gradually gain pilot proficiency flying low and slow over level terrain, on calm days.

A second goal for Moody was to make low-cost, completely portable ultralight flying available to the average builder, using an aircraft which one man could handle, transport, assemble and teach himself to fly anywhere with relative safety. The powered Icarus II, which Moody named *Easy Riser*, was a unique configuration among

Fig. 1-1. Jack Lambie revived the biplane hang glider movement in America with Hang Loose, built for $24.95.

rigid-wing ultralights. It had a parachuting type stall, important to low-time pilots winning their wings in foot-launched craft.

Easy Riser's parachuting capability comes from the wing design, which includes a tip washout to permit the outer panels to keep flying when the middle section stalls. This was not a new idea, but one that acts as a sort of air brake to hold the sink rate down. Stability is further enhanced in both pitch and roll axes by a combination of dihedral and sweepback.

The wingtip rudders, mounted on ball bearings for smooth operation, yaw the wing toward the direction in which the rudder is deflected by creating tip drag. This arrangement was developed by Taras Kiceniuk in his Icarus II and Icarus V hang gliders. Easy Riser emerged from Moody's conception five pounds lighter than Icarus II, Moody claims it is stronger, with higher lift-drag ratio, better penetration, lower sink rate and faster climb under power. It can be assembled in only six minutes.

One of the more popular PHGs flying today, Easy Riser is available in kit form from Ultralight Flying Machines, PO Box 59, Cupertino, CA 95014, or you can order them through Moody at UFM of Wisconsin, Box 21867, Milwaukee, WI 53221. According to UFM of California, Easy Riser is a "third-generation" high performance, rigid-wing PHG, a production version of the earlier Demoiselle. They describe it as actually a modified Monowing—a main wing with an aspect ratio of 8.8 and a newly-designed, high-lift airfoil consisting of the former Monowing tail moved forward to become the staggered lower wing of the biplane configuration.

Call them what you will, Easy Risers are flying by the hundreds. Kits come ready to assemble with Mac 101 engines equipped with a compression relief valve and recoil pull starter, intake stack, exhaust muffler, in-flight pull starter rope and pulley, Lord motor mounts, ear plugs, prob hub and face plate, and epoxied propeller.

UFM also supplies a twist-lock throttle cable, 5-quart fuel tank and shutoff valve, fuel line, fuel filter, mixture adjustment control and primer bulb, with a three-position ignition switch, wires, terminals, safety switch with retractile wire, and all required paperwork.

The unique ignition switch consists of a three-position mouth-held safety switch that lets you select engine-off, engine-on, and engine-on-via-safety-switch modes. During takeoff and landing maneuvers the mouth switch is used, so the pilot can kill the engine simply by opening his mouth.

Ear plugs are a must because the noisy McCulloch engine is right behind the pilot's head. Moody found that the best propeller was a 28/8.5 blade capable of producing 65 pounds of thrust from the Mac engine at 7300 rpm. The pilot sits in a swing seat or harness. However, during early flights Moody suggests hanging from the hang tubes under the armpits and staying below 15 feet altitude while you become proficient in handling the craft. Moody liked a commercial seat manufactured under the name Safety-Pro Harness, by Aero Float Flights, Box 1155, Battle Creek, MI 49016. He felt this seat gave a better sense of security than a swing seat made with a board suspended by two ropes. However, Moody warns that the quick release buckles could fly back and shatter the propeller blades if released prior to shutting down the engine.

Unplanned Aerobatics

There was some question about what would happen if a PHG somehow became inverted in flight. The question was answered dramatically at the August, 1976 EAA Fly-In at Oshkosh, WI, when Moody overdid a maneuver while showing off. But let him tell it:

"I overdid a wingover type maneuver which resulted in flying my ultralight right over onto its back. After several seconds of stable, level—though inverted flight—recovery was accomplished. But not until after doing three tight loops, or tumbles, around the lateral axis. I now believe that the unplanned aerobatics during the recovery would not have occurred if I had turned the engine off with the safety switch as soon as the recovery attempt was initiated.

"However, wanting to try only one thing at a time, I delayed turning the engine off. Apparently, my inability to stay in one spot in the aircraft, coupled with the thrust of the engine at full power, combined to allow the airplane to carry over in to a second and third loop. Once the engine was turned off, the aircraft immediately recovered into level flight. There was no structural damage of any kind, although I am sure God played the major role in my being here to write this account of what happened. I am also now more confident than ever in the structural and aerodynamic integrity of these aircraft."

Moody, in admitting pilot error, absolved the Easy Riser of blame, as did EAA officials. Moody was permitted to continue flight demonstrations during the remainder of the event. It was a hair-raiser for sure—I was there and I held my breath as Moody's machine did what appeared to be a series of Lomcevaks down below treetop level where he disappeared from view. Moody has had a few other tight situations, including a landing in four feet of water when his engine lost power and prevented him from reaching shore during the 1975 Frankfort Hang Gliding Meet.

Introduction to the Public

Moody's appearance at Oshkosh in 1976 was a unique and thrilling introduction of the sport of PHG flying to the public and it was a harbinger of things to come. In 1977, there were two PHGs at Oshkosh and in 1978 a total of 25 powered ultralights appeared. They had a special field all their own set aside to keep them out of the way of the faster conventional homebuilts flying past the reviewing area.

Later on you'll read about the incredible scene at Oshkosh '79, when ultralights really turned out en masse and drew proportionate acclaim. Many of the new PHGs are not simply converted hang gliders, but aircraft designed from the start to provide maximum performance from minimum horsepower with a total weight well under 500 pounds.

New engines also are making the scene, in addition to the standard Mac 101s and the Chrysler/West Bend 820 converted outboard powerplants. Chainsaw engines like the 5½-horsepower Swedish A. B. Partners can be mounted in twin-engine configurations as marketed by Ed Sweeney's Gemini International, 655 Juniper Hill Road, Reno, NV 89509.

Another, larger engine now being used with motorgliders is the 24-horsepower, twin-opposed, two-cycle Dyad engine that turns up at 7500 rpm and was initially designed for use in remotely piloted vehicles. It weights a mere 12½ pounds. Then there's John Chotia's 456-cc single-cylinder Chotia 460 engine he designed especially for his Weedhopper PHG. It is able to put out 18½ horsepower. It replaced the Weedhopper's earlier converted 22-horsepower Yamaha motorcycle engine.

Aside from small, light powerplants, PHGs also need special aerodynamic qualities if they are to realize their ultimate potential. Only recently have our national research efforts come face to face with the problems of designing craft to fly at low Reynolds numbers for aircraft like the NASA Mini-Sniffer, designed to become the first airplane to fly through the thin atmosphere of Mars.

Chapter 2

The View From Australia

Australia was the scene of hang gliding's appearance as a sport, but few know that ultralight aircraft also were "born again" in that country. We are indebted to a Quantas Airlines pilot, Gary Kimberley, of 73 Queens Road, Connels Point, NSW 2221, for this report on "minimum aircraft" development there.

Kimberley, 1979 president of the Minimum Aircraft Federation of Australia, manufactures the delightful Sky-Rider Ultra-Light which represents a new approach to the problem of producing an ultra-cheap, ultra-safe sport aircraft. Though classified as a powered hang glider, Sky-Rider (Figs. 2-1, 2-2, 2-3) has the controls and handling characteristics of a conventional light aircraft. It is equally suited to the novice flyer wishing to progress to conventional light aircraft piloting or to the experienced light aircraft pilot who wants to step down to ultralights as a low cost sport or hobby. Sky-Rider is designed to be dismantled and folded for car-top transport and garage stowage. It is built of cable-braced aluminum tubing and Dacron covering as a high-wing monoplane, using a modified McCulloch 101 engine of 12 horsepower.

Sky-Rider's wingspan is 32 feet, the length is 18 feet and the height is 7 feet 10 inches. Empty weight is 210 pounds and maximum takeoff weight is 400 pounds. Cruising speed is 40 mph and it takes off and lands at approximately 20 mph, stalling in ground effect at 18 mph. It climbs at 150 feet per minute. As a recreational machine, no pilot license or aircraft registration is required in Australia under Air Navigation Order 95.10.

Fig. 2-1. Gary Kimberley, president of the Minimum Aircraft Federation of Australia, flies the ultralight Sky Rider which he designed.

Kimberley has been flight-testing a brand-new powerplant, his experimental alcohol engine, that might deliver twice the power of the McCulloch 101, complete with dual ignition and quiet, smooth operation.

Australia, says Kimberley, has many excellent hang gliding sites, mainly on the East Coast. The best and most popular, he says, is probably Stanwell Tops, just south of Sydney. While Easy Risers currently are the most popular type PHG's, Kimberley has an interesting story to tell about the rebirth of the sport.

On 28 March 1979, Kimberley gave a lecture on "Minimum Aircraft—Sport Flying For the Layman." It is reprinted with his permission:

To do justice to the history of our movement, I feel one should start with the experiments of Otto Lilienthal (Fig. 2-4) in Germany in the late 1890's. In my opinion, Lilienthal was the real father of flight. He was the first to achieve successful, controlled heavier-than-air flights. He was the first to be able to remain airborne for sufficient periods of time, to be able to establish satisfactory methods of control and to begin accumulating useful aerodynamic data. Prior to his tragic death in one of his later model hang-gliders, Lilienthal had thrilled the world with his daring feats and there is no doubt that his amazing exploits and remarkable successes provided the inspiration to Wilbur Wright to embark on his ambitious project to become the first to achieve powered heavier-than-air flight.

The First Minimum Aircraft

Using Lilienthal's tables of air pressures and "vaulted" (cambered) wings, Wilbur devised his first series of man-carrying gliders

Fig. 2-2. The cockpit area of the Kimberley Sky Rider is utter simplicity.

for his Kitty Hawk experiments. He shrewdly realized, however, that weight-shift control was far too limiting and that successful powered flight would depend on the development of a satisfactory method of aerodynamic control.

The breakthrough came when Wilbur discovered the wing-warping technique which enabled him to achieve fully-controlled banked turns. With the addition of a workable engine and propulsion system, the first successful powered airplane was born—the world's first minimum aircraft.

The rapid progress of aeronautics through this embryonic, ultra-low and slow stage was so swift that very little detailed research was done in this area. The quest was for bigger, more sophisticated aircraft of ever-higher performance. Two world wars saw quantum jumps in the development of aviation that left the

Fig. 2-3. Clean, simple lines of the Sky Rider show here. Note the wide ailerons.

minimum aircraft regime far behind. The result was that it became an abandoned and forgotten art.

With today's enormously costly and sophisticated aircraft, research and development is carried out in infinite detail and its cost is measured in thousands of millions of dollars. Publications are available to the modern aircraft designer that can provide him with aerodynamic data to the Nth degree, or just about any aerodynamic shape he likes to think of, in the conventional flight regime. But if you want to design and build your own minimum aircraft with an all-up weight of less than 400 pounds, a single-surface wing with a normal operating speed range of from 15 to 35 knots, then you're on your own!

The Rediscovery of Hang Gliding

Some flying had been done by water-skiers using flat pentagonal kites behind powerful speed boats, but it was the invention of the Rogallo Wing by Dr. Francis Rogallo and its later development for the United States space program, that really revolutionized the sport and brought about the hang-gliding boom that swept around the world in the late 1960s and early 1970s.

The *Rogallo* (Fig. 2-5) or *para-sail* was designed as a flexible delta wing which could be rolled up and folded, and carried on a space vehicle in such a way that, after reentry, the Rogallo could be popped out like a parachute—enabling the spacecraft to glide down to a controlled landing on the ground. As it turned out, NASA's requirement for the Rogallo never developed. But the tow-kiting

Fig. 2-4. Otto Lilienthal was a pioneer hang glider pilot. He was killed when he attempted flight in a powered glider.

Fig. 2-5. An early Rogallo wing hang glider with the pilot riding in the prone position.

enthusiasts were quick to see the potential of the device in their application and so a brand new sport was born—the sport of hang gliding.

Much of the early pioneering in hang gliding was done in Australia by John Dickinson, Bill Bennett and Bill Moyes in Sydney. The breakthrough occurred with their introduction of the A-frame control bar, which enabled complete pilot control of the kite in free flight. This meant that, on attaining a safe height, the pilot could release from the tow rope and glide down to a landing wherever he chose—with the kite under full control.

It wasn't long before others, such as Steve Cohen, began experimenting with tethered flying in strong sea-breezes, gliding down sand dunes at Kurnell and ridge soaring at Stanwell Park, in the shadow of Lawrence Hargrave, pioneer Australian aircraft designer.

Powered Hang Gliders

After visits to America by Bill Bennett and Bill Moyes, the sport of hang gliding really began to boom. It wasn't long before the Americans, with their flair for innovation and ingenuity, were producing hang gliders of increasingly greater sophistication and higher and higher performance capabilities. The Rogallo's nose angle was widened, aspect ratios were increased and the sails were progressively flattened by decreasing the degree of billow.

As the Rogallos became more and more like flying wings and began to reach the limits of their design potential, the next step became obvious—the development of the Rigid Wing. It wasn't long before such sophisticated gliders as the Icarus, Easy Riser, and

Mitchell Wing began to appear on the scene. Although they were heavier and lacked the portability and quick assembly and breakdown times of the flexible wing kites, they more than made up for this with their superior performance capabilities. The incorporation of wingtip rudders or spoilerons became essential on these high aspect ratio gliders. Weight-shift alone was not enough to provide adequate roll control. Instrumentation began to appear in the form of airspeed indicators, altimeters and variometers. Naturally, it was to be only a matter of time before someone got the bright idea of bolting a chainsaw engine onto a hang glider.

In 1974 Ron Wheeler, a Sydney boat manufacturer, who had begun making hang gliders in his Carlton factory, designed a revolutionary new high aspect ratio, flexible wing glider, with a conventional aircraft-type configuration that had a performance capability approaching that of some of the rigid wings. This rather unusual hang glider was called the *Tweetie* and employed an ingenious method of using yacht masts and sails.

Although Tweetie never became a huge success as a hang glider, it made a vital contribution toward the birth of the minimum aircraft movement through the ready adaptability of its design for conversion into a very basic, powered miniature airplane. The *Skycraft Scout*, the first of the true, modern-day minimum aircraft was born (Figs. 2-6, 2-7 and 2-8).

Air Navigation Order 95.10

Having produced the world's first viable minimum aircraft, it then became necessary for Ron Wheeler of Skycraft Pty. Ltd. and his colleague Cec Anderson to obtain official approval from the Australian Department of Transport in order to fly it legally. In November 1976, permission was finally granted by the Department in the form of an Air Navigation Order which exempted this class of aircraft from all normal requirements of Air Navigation Regulations but imposed strict limitations on their operation.

This Air Navigation Order, ANO 95.10, was an enormous breakthrough. It enabled the development of an entirely new branch of aviation and brought about the birth of an exciting new sport—minimum aircraft flying. It enabled personal, recreational flying to be brought back within the reach of the ordinary private citizen and displayed an unusually generous and progressive outlook on the part of the DOT. It was also a tremendous breakthrough for Ron Wheeler and Cec Anderson of course. It now meant that their Skycraft company could legally manufacture and sell their Scouts in

Fig. 2-6. Ron Wheeler's Australian Skycraft Scout was the first of modern day minimum aircraft.

the minimum aircraft category. Almost overnight their sales began to boom. There are now more Scouts flying than any other minimum aircraft type anywhere in the world.

Now, of course, there are a number of other designs also flying (Fig. 2-9), such as Col Winton's Grasshopper, (Fig. 2-10) which is also in production in Sydney, Steve Cohen's Ultra-Light which is currently under development and my own experimental design, the Sky-Rider. This, however, is just the beginning. The field is wide open to personal enterprise, and individual, inventive ingenuity.

Designing a Minimum Aircraft

Imagine that in a sudden fit of creative enthusiasm you have decided to design and build a minimum aircraft of your very own creation. This is not as far-fetched as it sounds, and is probably quite

Fig. 2-7. A Skycraft Scout fitted with pontoons flies nicely.

Fig. 2-8. Ron Wheeler's Skycraft Scout appeared at the EAA Fly-In at Oshkosh in 1979.

within the capabilities of most of us if we are prepared to put our minds to it. However, before we start getting too involved in the design and construction of our minimum aircraft, we must really make sure that we know exactly what is involved in the ANO. There is not much point in spending a lot of time, effort and money on an airplane that could never be legally flown.

Air Navigation Order 95.10 applies to power-driven, heavier-than-air, fixed-wing aircraft having a maximum takeoff weight not exceeding 400 pounds and a wing loading not exceeding 4 pounds per square foot. It therefore includes both powered hang gliders and minimum aircraft and these machines are thereby exempted from all the normal requirements applicable to conventional aircraft.

It sounds good—no pilot's license, no registration, no Air Navigation Charges. But the sting is in the tail—the operational limitations imposed, conditions which can briefly be summarized as follows.

Aircraft to which this ANO applies shall not, under any circumstances, be flown:

- In cloud.
- At night.
- Over built-up areas.
- In instrument-flying conditions.
- In commercial operations.
- In any aerobatic type maneuvers.

In addition, except with written permission of the regional director of the Department of Transport, these aircraft shall not be flown:

Fig. 2-9. The Hovey Whing Ding from America is popular in Australia. This one was built by David Ecclestone.

- At a height in excess of 300 feet above terrain.
- Within 5 kilometers of a government or licensed aerodrome.
- Within controlled airspace.
- Within any prohibited or restricted area.
- Within 100 meters of members of the public.
- Within 100 meters of any building.
- Within 300 meters of any sealed road.
- At any regatta, race meeting or public gathering.

Virtually, what all this means is that if you want to fly minimum aircraft you will have to get out into the country, away from any towns or built-up areas, so that if you have a prang you're not going to hurt anyone but yourself. The Department of Transport's concern, of course, is public safety.

Fig. 2-10. The Winton Grasshopper is another Australian ultralight.

If, after all this, you still want to go ahead with the design of your minimum airplane, the key limitation will be the 400 pound minimum takeoff weight (MTOW). If you can design your aircraft to comply with that, it is fairly unlikely that you will exceed the 4 pounds per square foot wing loading limit.

The biggest single problem facing minimum aircraft designers at the moment is that of finding a suitable engine. It must be extremely light in weight (preferably under 30 pounds), put out a minimum of around 10 to 12 horsepower, be completely reliable and readily available at a reasonable cost. It must also be decided how your aircraft is to be transported to and from the flying site—on the car roof-rack, on a trailer or towed.

Next, it must be decided whether it's going to be a tractor or a pusher, enclosed or open cockpit. And finally, all the nitty-gritty decisions on how you're going to get it all together in such a way that it will actually work, while still remaining inside the legal weight limit. And I can assure you, "It ain't easy!" If you finally get your masterpiece finished, that alone will be quite an achievement. Statistics show that, of those who start homebuilt aircraft projects, very few ever actually finish them.

Flying Minimum Aircraft

So, having paid out the money, worked all those hours and solved all those seemingly innumerable problems one by one, you are now the proud owner of a brand new minimum airplane of your very own design. It is a thing of beauty, a unique work of art, the end result of your labor of love and tangible proof of your creative genius. All you have to do now is learn how to fly it without "writing it off" the very first time you try to get it into the air! If you can do that, you have done very well indeed. You will experience the pride of achievement and a degree of personal satisfaction enjoyed by few others—a handsome reward which you will have well and truly earned.

A word of warning: flying minimum aircraft is a new and specialized art. It can only be learned through practice and experience and should be approached cautiously and conservatively, advancing only one small step at a time. It presents many problems not encountered in other branches of aviation and, unfortunately, it provides a fertile breeding ground for over-confidence and carelessness.

Previous light aircraft flying experience, while helpful, will not provide a pilot with the skills necessary for flying minimum aircraft.

In fact, in some instances, such as in the case of the Skycraft Scout, previous conventional light aircraft experience can be a disadvantage because of the different control system and the different flying techniques required. The experienced light aircraft pilot has the added burden of endeavoring to overcome his normal instincts and habits built up over many hours of conventional aircraft flying.

We have had a number of minimum aircraft accidents and incidents recently, involving licensed light aircraft pilots. In each case, except one, the pilot was very lucky to have escaped serious injury. The exception, unfortunately, was fatal.

Minimum aircraft operate at low Reynolds numbers, dimensions are small, weights are virtually in the hang-glider category and airspeeds are extremely low—with very little margin above the stall. The weight of the pilot is usually around 50 percent or more of the total and can be a critical factor in the aircraft's performance.

Engine power is minimal and very few minimum aircraft could sustain a level turn with more than about 15 or 20 degrees of bank. Even when climbing with full power, a minimum aircraft can easily begin to lose height if it suddenly encounters a downdraft or runs into an area of sink.

Large control inputs can create such an increase in drag that the aircraft will be forced to descend to maintain airspeed. Wind gusts which would not even be noticed in a light aircraft can easily be 15 or 20 percent of the flying speed of a minimum aircraft. For this reason, minimum aircraft should never be flown in gusty conditions or in winds over about 10 or 12 knots maximum.

The Minimum Aircraft Federation of Australia

In granting the exemption under ANO 95.10, the Department of Transport expressed the wish that some sort of controlling body be formed. Ron Wheeler subsequently wrote to all Scout owners asking if they would be interested in forming a Minimum Aircraft Association.

On April 20, 1978, a meeting of interested persons was held at the Royal Aero Club of NSW in Bankstown, where Nicholas Meyer, a Sydney businessman and Scout owner and flyer, became president of The Minimum Aircraft Federation of Australia. The organization is growing steadily with members in every state of Australia plus a number overseas. The aims and objectives of AMFA are:

- To promote minimum aircraft flying as a safe, low-cost sport that will bring enjoyment of flying back within the reach of the ordinary citizen.

- To protect the right of the private citizen to build and fly his personal aircraft within requirements of the law.
- To safeguard the interests of minimum aircraft enthusiasts throughout Australia.
- To guide and control development of the sport in an organized and constructive manner, with safety as the prime objective.
- To encourage formation of minimum aircraft clubs throughout Australia and offer them guidance and assistance and seek their affiliation with the Federation.
- To act as a central communications and coordinating body for the minimum aircraft movement as a whole, to liasie with the Department of Transport and act on behalf of the members where necessary.
- To ensure that costs, paperwork, and rules and regulations are kept to a minimum commensurate with safety and do not become the inhibiting burden that they have become with conventional aviation.
- To promote friendship, courtesy, mutual assistance, and the return of a spirit of chivalry in the air.

The key aim is, of course, to preserve ANO 95.10, for if we should ever lose that, we're out of business.

The Current Situation

In addition to Scouts, we now have Grasshoppers, Whing Dings, Sun Funs, and one Sky-Rider. We are also taking an interest in such powered hang gliders as the Easy Riser, Quicksilver and Mitchell Wing, as they operate under ANO 95.10 and are now being fitted with wheeled undercarriages and in some cases additional aerodynamic controls.

For purposes of definition, we are working on the principle that if the pilot hangs suspended in the aircraft and weight-shift is used in all or any one of the three control axes, then it is a hang glider. If it flies like an airplane with all aerodynamic flying controls, then it is a minimum aircraft. However, some powered hang gliders, such as the Easy Riser biplane, when fitted with its tricycle undercarriage, is near enough for our purposes to be considered a minimum aircraft.

A Giant Step Backwards. The minimum airplane has been jokingly described as a "giant step backwards." This rather hilarious and astute comment is really quite true. Minimum aircraft flying represents a return to the good old days of early aviation when flying machines were very basic and aviators flew them for the sheer thrill of flying.

The years that have elapsed since Wilbur and Orville Wright made their first short flights over the sands of Kitty Hawk have seen aviation progress from an exciting experiment to the world's most automated and regulated mode of transport. Now, by going back to basics, we are just beginning to rediscover what it's really like to fly!

The Future. The advent of the minimum airplane has made it possible once again for the ordinary citizen to own and fly his personal aircraft. It brought about the discovery of an entirely new and exciting branch of aviation, halfway between the hang glider and the old "ultra-light" aircraft.

Conventional private flying is becoming so bogged down with rules, regulations and red tape, and is going to become so prohibitively expensive over the next few years, that our minimum aircraft type of flying will probably eventually become the only way in which an ordinary citizen will be able to enjoy the privilege of personal, recreational flight.

Our sport is still very much in its infancy, but I believe we are on the verge of an upsurge in sport flying that will sweep throughout Australia and right on around the world. Minimum aircraft flying has the potential to become the world's largest sport aviation movement, provided it is handled correctly and is not fouled up in the early stages of its development.

Already in Sydney alone there are at least four different locally-designed minimum aircraft, either already in production or in advanced stages of development, all of which will be available to Australian enthusiasts at a price well within the reach of ordinary wage earners. At the moment, we are well ahead of the rest of the world in the minimum aircraft field, including the United States, thanks mainly to ANO 95.10. Just how long we will be able to maintain our lead will be interesting to see, but obviously it will depend to a great extent on our own enterprise and enthusiasm.

Conclusion. The Minimum Aircraft Federation was formed to further the aims and safeguard the interests of enthusiasts throughout Australia who simply wish to fly safe, simple aircraft which they can afford to buy and afford to fly for the sheer enjoyment of flying. People might ask: "What is it about flight that makes so many people strive so hard to achieve it?"

Perhaps the best answer was given by Wilbur Wright back in 1905: "When you know, after the first few minutes, that the whole mechanism is working perfectly, the sensation is so keenly delightful as to be almost beyond description...More than anything else, the sensation is one of perfect peace, mingled with an excitement that strains every nerve to the utmost—if you can conceive of such a combination."

Australians on the Go

To amplify Gary Kimberley's account of Minimum Aircraft flying in Australia, let's look back across the years to the turn of the century when Lawrence Hargrave made an important contribution to the young science of aerodynamics, in its slow evolution from kites to airplanes.

In 1892, Hargrave sent a scientific paper to Octave Chanute, in America, to present at the world's first conference on aerial navigation at the Chicago World's Fair. In it he described how he had built kites with square sides—not just flat plates or single curved surfaces. He called them box kites, because they looked like flying boxes.

Hargrave used curved surfaces on the top and bottom of his box kites, but he added something even more important to their basic design. Using two surfaces instead of one, he doubled the force of lift. In addition, the sides of his box kites acted like rudders to keep the kites flying straight.

Hargrave learned two important things from his box kites. First, he found that their curved surfaces developed twice as much lift as flat surfaces. And if he flew a box kite at a relative angle (to the wind) of about 45 degrees, lift and drag were about equal. He then tried flying one tilted a little lower and discovered that lift exceeded drag considerably.

Later on, Chanute, Lilienthal and the Wright brothers would refine the study of the ratio of lift to drag, called LD, and compile detailed tables for wing curves of different shapes. Another Australian, Richard Pearse, developed a powered flying machine of his original design that reportedly flew on several occasions—though not too well. Pearse's experiments actually were carried out in New Zealand, but Australia considers him a sort of adopted son to show the world that aeronautic inventiveness flourished years ago "down under."

With the current rebirth of interest in ultralight flying in Australia, new designs are appearing with regularity. However, several American designs already are highly popular in the Outback of Australia, where sheep ranchers use them to survey their flocks from the air.

Among the American imports that have gained wide acceptance in Australia is Bob Hovey's little biplane, the Whing Ding, perhaps the world's smallest ultralight biplane that weighs only 123 pounds empty and can attain a speed of 50 miles an hour. Hovey has sold more than 6000 sets of Whing Ding plans around the world at $20 a set, not bad for an amateur designer! In a later chapter, the Whing Ding and Hovey's latest design, the monoplane Beta Bird, are discussed further.

As Kimberley points out, the Volmer VJ-24 Sun Fun and John Moody's Easy Riser also are highly popular imports in Australia. As the sport of Minimum Aircraft flying has caught hold there, some highly interesting local designs have shown up.

The Hornet

One of the newer models is David Betteridge's flying wing minimum airplane, the Hornet 160. The pilot sits in an enclosed cockpit atop the swept-back wing, whose span is 33 feet and area 162 square feet. Wingtip rudders are operated independently by the pilot's feet for turning, or together to serve as dive brakes.

Betteridge was an aeronautical engineer with Hawker de Havilland Australia in Sydney, then joined Free Flight Hang Gliders Proprietary Limited of Adelaide, South Australia, to develop a powered hang glider. This development work resulted in the Hornet 160 design, similar to the American Mitchell Wing, except for the higher pilot position.

The Hornet 160's powerplant is a modified 177cc two-stroke motorcycle engine mounted behind the cockpit, driving a ducted fan. The craft's top speed is 70 mph, its climb rate is 400 feet per minute and its stalling speed is 25 mph. Stressed to 6 G's, the machine can handle severe turbulence. Construction is of aluminum alloy, plywood, polyurethane foam and fiberglass. The wing's leading edge is plywood-covered and the rest is covered with fabric. Assembly or disassembly takes only 10 minutes. At this writing, plans were firming up for a production run.

The Scout

I had the pleasure of watching a Skycraft Scout fly at the 1979 EAA Fly-In at Oshkosh, WI, and I admired its unusual grace and good performance. This confirmed what I read about it in Alan Chalkley's report from Australia to the British publication *Popular Flying* in their March-April 1979 issue.

Chalkley pointed out the unique way the designer, Ron Wheeler, a Sydney boat builder, had adapted the highly efficient sail of a racing dinghy to aeronautical use. The result was almost birdlike flexibility. Wheeler started off using a section of a dural boat mast for the leading edge of the wing, with lift and landing wire braces rigged to a cabane strut above and to the A-frame below. Single stay wires to the forward fuselage handle the drag loads.

The wing is covered with Dacron, stiffened with seven aluminum alloy ribs sewn into place in a manner permitting the covering to be rolled up for storage or cartop transport. The magic of the design becomes apparent in flight. As the angle of attack increases, the wing flexes progressively from the tips toward the root. This will effectively wash out the high incidence to prevent stalling. Chalkley reported witnessing a Scout flown by Ces Anderson, Wheeler's colleague, put into an extreme nose-high attitude without power at 150 feet above ground level and watching it gently settle back into landing attitude in bird-flight fashion.

Construction of the Skycraft Scout is simple. The fuselage is made from a Dural spar, beneath which is slung a tubular A-frame in which the pilot rides comfortably in a fiberglass seat. The main gear is suspended from a steel leaf spring beneath the forward end of the framework in a manner to absorb shock loads from hard landings and to prevent injury to the pilot.

There are no ailerons or wing-warping involved since the all-sail wing provides adequate lateral control. Rudder control is linked to the control stick which is moved sideways toward an intended turn, eliminating the need for foot rudder pedals. Your feet rest on a simple crossbar.

Wheeler's factory also produces the small Pixie powerplant for the Skycraft Scout, a two-stroke, single-cylinder affair mounted at the front end of the fuselage boom with rubber shock mounts. A geared chain drive with a 4:1 ratio reduction is automatically lubricated. Wheeler also went to the advanced solid state electronic ignition system.

A half-gallon fuel tank is mounted above and behind the Pixie engine and operates by gravity feed. This amount of fuel is good for 40 minutes flying, giving a range of 28 miles, no reserve. When the engine quits you simply glide down and land, providing you are over open country. Australians are lucky here, for there are thousands of square miles of open space for sport flying in minimum aircraft without penetrating controlled airspace or violating the restrictions of ANO 95.10, which is written to keep the little guys at a safe distance from airports and paved highways or buildings.

Chalkley reports that Wheeler maintains a pilot training area where clipped-wing Scouts run back and forth without taking off, in the manner of World War I Penguin trainers at Pau in France. Short hops with shallow-banked turns follow. Within two or three hours a neophyte pilot is considered ready to leave the nest. Veteran pilots of conventional aircraft seemed to have difficulty learning to handle the Scout without rudder pedals, but once they got the hang of it there were no problems.

Wheeler has manufactured and sold close to 300 Skycraft Scouts at this writing—mostly in Australia, though one was reportedly flying on Christmas Island. Sale price was quoted as $1885 (£1075).

The Scout weighs only 122 pounds empty or 297 pounds all-up. Its wingspan is 28 feet 6 inches and wing loading is a light 2.03 pounds per square foot. It can take off at 24 knots in about 200 feet, climb at 180 fpm, hit 42 knots and cruise at 36 knots.

Stall speed is 18 knots in ground effect and the stall is very gentle and straight forward due to its unusual wing design. It lands at 24 knots in a 90-foot rollout. Glide ratio is listed at 7:1, maximum pilot weight 175 pounds. The Pixie Aero engine displacement is 173cc.

On a visit to Australia by 45 Canadian and United States members of the Experimental Aircraft Association in the Spring of 1979, the ultralights displayed a real international flavor at a gathering of homebuilts at Bowral for the Sport Aircraft Association of Australia's Fifth National Fly-In convention. Among the wee ones that drew admired attention was an Australian design, the Grasshopper, with a streamlined cockpit area, midwing airfoil, tricycle gear and a tiny pusher engine pylon-mounted behind the pilot's head.

But if Australia is leading the way in Minimum Aircraft, the United States today remains the site of the largest concentration of ultralight aircraft. In the following chapters, you'll read more about the made-in-America brands.

Chapter 3

An American Tradition

It should come as no surprise that many of the most recent advancements in design of ultralight flying machines have their roots in antiquity—primarily in methods of achieving stability and control. Well known is the pioneer work of such aeronautical innovators as Sir George Cayley in England, Adolphe Penaud in France, Otto Lilienthal in Germany and the Wright brothers in America, who were first to produce and master a workable heavier-than-air flying machine.

Less known is the work of countless experimenters at home and abroad, who independently strove to duplicate the mechanisms of bird flight and largely failed, primarily because they did not quite understand how to separate the two main functions of the bird's wing—to provide both lift and propulsion.

An Unusual Homebuilt

Orville Wright was just 5 years old when a forgotten American aviation pioneer was making local history in a small Tennessee town with an unusual homebuilt ultralight machine. In 1876, Melville Milton Murrell was the 21-year-old son of the postmaster at Panther Springs, a waystop on the Knoxville stagecoach road, who like so many others, turned to birds for inspiration in aircraft design.

Murrell settled at length on an ornithopter (wing-flapper) design. He built a set of slatted wings resembling a venetian blind (Fig. 3-1). The slats opened on the upstroke and closed on the downstroke. When they were held rigidly flat they formed a sustain-

Fig. 3-1. Melville Murrell was flying gliders like this more than 100 years ago in Panther Springs, Tennessee.

ing plane that made it a reasonably good glider. Today, wing slats are occasionally incorporated to maintain a laminar airflow over a wing at high angles of attack. But to young Melville they were simply imitations of a bird's feathers, which, we now know, actually do serve as laminar flow devices. Observant of wheeling hawks, he also was aware that birds often fly for long periods without flapping. One reason why he incorporated in his design its convertible nature was to provide both lift and propulsion.

After several failures, Murrell on December 4, 1876, was able to write to a friend, Will Turner, that he had mailed a model of his machine to the United States Patent Office. "I can say 'Eureka! Eureka!' For it works like a charm!" he exclaimed. In 1877, the Patent Office saw fit to grant Murrell a patent, No. 194,104, in a description of which Murrell explained: "The machine is operated by the hands and feet" and was "guided by the tail" to which a "wriggling or partially rotary motion is imparted."

An historic photograph, given to me by a member of the Murrell family, shows the machine at the Murrell farm, with several witnesses who had been summoned to attest to its successful operation. Witnesses were the Reverend W. C. Hale, John Mathis, Henry Mullins, and F. Roger Miller, who later would become president of the United States Chamber of Commerce.

Family sources report that, after a few glides across the family apple orchard, young Murrell—on the advice of his father—turned down an offer of $60,000 for his patent. But Melville was simply ahead of his time. In 1876, German inventor Nikolaus Otto had just developed the first internal combustion engine that operated on natural gas. Shortly thereafter, Gottlieb Daimler would convert it to burn liquid gasoline.

Stability and control problems had not been solved by the Murrell flying machine and young Melville finally gave up flying and turned to follow the sawdust trail.

He preached the gospel for the next 45 years as a Methodist circuit rider. His early dreams of flying were carried in his heart and are cited here primarily because they represented dreams shared by so many other early American inventors whose ideas found their way into the Patent Office files.

Octave Chanute

In 1894, the American civil engineer and bridge builder Octave Chanute published many of these early concepts in a book, *Progress in Flying Machines*, and followed up on his research with the design and construction of a rigid biplane hang glider whose upper and lower wings were stiffened by the Pratt truss design common to railroad bridge engineering practice.

Although Chanute was too old to fly his gliders except on rare occasions, an assistant, A. M. Herring, did make numerous glides in them. He maintained longitudinal balance with body shift in the manner of modern-day hang glider pilots. The problem was to keep the center of lift of the wings coincident with the craft's center of gravity. Chanute used a horizontal tail surface, as tried out by Penaud in France, to more easily maintain longitudinal stability in gusty air.

After reading Chanute's book, Wilbur and Orville Wright contacted Chanute in 1899 and settled on the Chanute biplane configuration for their early glider designs. There was a major difference. The horizontal plane was placed in front of the wings in canard fashion, a design feature recently revived by Burt Rutan in his highly successful VariEze design. The Wrights also developed an innovative method for maintaining lateral control by adding a wingtip-warping control system inspired by the "wingtip torsion" they observed watching buzzards in wheeling flight.

Alexander Graham Bell

Independent of the Wright development program, the Aerial Experiment Association, a group of enthusiasts organized by Dr. Alexander Graham Bell, struggled with the problem of stability control in flight. While riding a train one day, Dr. Bell conceived of the aileron as a device for changing the relative angle of attack, using a mid-wing control surface hung between top and lower wings of the AEA's biplane.

Dr. Bell obtained a United States patent on the aileron system, in the name of all members of the AEA, and later it formed the basis for a drawn out court battle over priority for invention of a controllable flying machine. The Wrights claimed that their wing-warping patent, with a vertical rudder to offset adverse yaw, was basic to the three-torque control system of aircraft management in pitch, roll and yaw. A patent truce was finally arranged at the outset of World War I to allow manufacturers to get on with the business of building warplanes.

The problem of adverse yaw and how to control it stumped the experts at first. They sought to attain control in turns by increasing the angle of attack of the outer wing to make it rise into the required bank. In so doing, the increase in angle of attack at the same time was accompanied by an increase in drag that tended to pull the rising outer wing backward.

The Wright Brothers

To solve the problem, the Wrights improved on the birds, so to speak, by adding vertical fins behind the wing on their 1902 experimental biplane glider. It worked at first, until steep turns were attempted and the machine developed a tendency to overbank. In correcting for this with opposite wing warping—to increase the angle of attack of the inner, lower wing and make it rise—the inner wing in the turn, which was traveling slower than the outer wing, reached a stall. A helical dive ensued and started the deadly stall-spin maneuver that has killed so many pilots since.

Orville puzzled over this behavior and finally decided they should abandon the double fixed rudder in favor of a single, movable rudder. Wilbur came up with the idea of linking the rudder and wing-warping controls together to eliminate the need for pilot coordination in working two levers at the same time. This linkage was similar to the famous contribution of Fred Weick in designing the popular Ercoupe so that it could be flown without rudder pedals.

The Wrights settled on independent pitch, roll and yaw control, abandoning the rudderwing warping linkage in their 1904 machine (Fig. 3-2) for a method of three-axis control most widely in use today. The wing-warping was done with a novel hip saddle. The pilot merely moved his tail left or right in the direction of intended turn. It seemed like a natural progression from control by body shift used by Lilenthal and Chanute, but would eventually be replaced by the standard "joystick" and rudder bar for three-torque control.

The mystery of the wing curve was another feature of the airplane that puzzled early experimenters unfamiliar with fluid dynamics. While the Wrights attained success by means of their well-known series of wind-tunnel tests on various airfoil shapes, most of their contemporaries simply tried to adapt the curves of the wings of soaring birds. The Wrights also proved to themselves in their small wind tunnel that a long, narrow wing of high aspect ratio worked better than a short stubby wing of wider chord—the way nature designed buzzards and soaring seabirds.

The Secret of Flight

One so-called inventor, presented a paper at the 1893 International Conference on Aerial Navigation at the Chicago World's Fair claiming that the secret of flight lay in goose feathers. Serious investigators, like Chanute, Samuel Pierpont Langley and Albert Zahm, who had organized the meeting, listened with something less than awe when the delegate climbed to the podium and asked rhetorically:

"How can a wild goose carry itself so easily? Weight every feather and they will not total one pound. Yet, pick those feathers off goose and he can come no nearer to flying than we can!"

Alternately cheered and booed, he paused, then continued: "Thus it is clearly demonstrated that one pound of goose feathers can pick up 19 pounds of goose and carry it through the air at half a mile a minute!" All that was needed, he concluded, was someone to discover the secret of goose feathers and the problem of flight would be solved.

Not all the delegates to the 1893 Chicago meeting were so wild in their imaginations. Delegates came or sent papers with such practical ideas that Chanute was moved to remark that the airplane had already been invented and it only remained to get all the ideas together and make them work. From California, John J. Montgomery had built and flown an ultralight glider on the slopes of Otay Mesa near San Diego. Others were on the right track, but it took Orville and Wilbur to assemble the first powered heavier-than-air craft and learn to fly it.

As in the past, today young inventive people are making new discoveries in the area of ultralight flight. They are finding new ways to make their craft inherently stable, or more controllable in the realm of low-speed flight, where airflow is measured in low

Fig. 3-2. Wilbur Wright made the first powered flight in America on December 14, 1903.

Reynolds numbers. Others are rediscovering basic laws of aerodynamics put aside long ago in the rush to fly faster, higher and farther.

Demoiselle

The first decade of this century saw a proliferation of small, light, powered flying machines take to the sky—among them Alberto Santos Dumont's pretty Demoiselle, (Fig. 3-3) also called Le Santos No. 20. In 1909 it attracted much attention at a Paris air exhibition. It measured only 18 feet from wingtip to wingtip, half the span of a J-3 Piper Cub, and weighed only 242 pounds. It was constructed from bamboo and muslin. Its two-cylinder engine developed an amazing 30 horsepower and swung a fat-bladed propeller

Fig. 3-3. The Santos Dumont Demoiselle was the first plans-built homebuilt ultralight. This one was built recently by Earl M. Adkisson of Atwood, Illinois.

6 feet 6 inches in diameter. It was so heavy that it served as a flywheel and produced huge gyroscopic loads that made the machine difficult to control in turns. Santos-Dumont rigged the controls with an elevator lever and a control wheel hooked up to the rudder.

Homebuilders flooded Santos-Dumont with requests for plans for the Demoiselle. In 1910, *Popular Mechanics* magazine offered sets of working drawings for $2 a set, announcing: "The machine is unencumbered by patent rights, the famous aviator preferring to place his invention at the disposal of the world in the interest of the art to which he has devoted his life."

By contrast, the Wright Kitty Hawk biplane was jealously guarded to prevent others from copying the design. So deeply were Orville and Wilbur involved in litigation in 1916 that Congress, facing involvement in World War I, arranged a patent truce and appropriated $640 million to initiate a warplane production program.

Slow starting as a result of the Wright litigation, the United States lagged far behind Europe in production of military aircraft. When the Armistice was signed November 11, 1918, America had only 757 pilots and 481 observers, along with 740 planes and 77 observation balloons at the front. However, thousands of brand-new crated Standards and JN4D trainers flooded the postwar market. The glut of surplus aircraft sorely impeded development and sale of new private aircraft.

The White Monoplane

The era of barnstorming gypsy flyers was born after the war. But at the same time there was a strong appeal in ultralight flying among returning service pilots who wanted to build and fly inexpensive machines such as the White Monoplane that first appeared on the market in 1917. The design was developed by George D. White, who offered plans for $2 a set.

"Think of flying with an ordinary twin-cylinder motorcycle engine!" White's advertisement suggested. "This is the only aeroplane that will do it. It is the smallest and most efficient of all aircraft. No longer is flying the sport of acrobats or millionaires. If you can use a hammer, saw and a pair of pliers, you can build one of these remarkable flyers for a few dollars and in spare time if necessary."

Like the early Wright machines, the White Monoplane used the canard design with its tail up front. While it never became popular, it preceded a number of other ultralight designs that appeared soon afterward.

The Penguin

In the September, 1919 issue of *Popular Mechanics*, the magazine again offered plans for an ultralight. The Penguin was patterned after a World War I primary flight trainer that appeared in 1916 while American Escadrille cadets were learning to fly at Pau, France. *Aviation Magazine* described its operation at the time:

"First of all, the student is put on what is called a roller. It is a low-powered machine with very small wings, and is strongly built to withstand the rough wear it gets, and it cannot leave the ground. The apparatus is known as the Penguin, both because of its humorous resemblance to the quaint Antarctic birds and its inability, in common with them, to do any flying.

"A student makes a few trips up and down the field in a double-control Penguin and learns how to steer with his feet. Then he gets into a single-seated one and tries to keep the Penguin in a straight line. The slightest mistake will send the machine skidding off to the right or left and sometimes, if the motor isn't stopped in time, over on its back. Something is always being broken on a Penguin, so a reserve flock is kept on hand."

A first-hand account of construction and flying a Penguin ultralight comes from Tom Gunderson, a veteran crop-duster and commercial pilot of Twin Valley, MIN, who in 1972 made the maiden flight in a Penguin built from the original plans he found in an old issue of *Popular Mechanics*.

It was in 1929, during the Great Depression, that Gunderson had built his first plane, a Pietenpol Scout, and taught himself to fly. It seemed a more practical machine than the Penguin, though the *Popular Mechanics* article had first turned him on to homebuilding. When World War II came along, Gunderson instructed cadets with the Civilian Pilot Training Program, then went on to crop-dusting after the shooting stopped.

Recently he recalled the ridicule he suffered when he first started on the Pietenpol and he mused: "The old boys who made fun of kids trying to fly are all dead now. I tell you, for a young man to start flying now is as hard as when we were young, back before the big war. The way I figure it, there's a million small fishing boats for recreation, but what we need is a small, cheap sports plane for fun and training."

It turned out that Gunderson's Penguin, which he finally got around to building, cost him less than $50, not including its snowmobile engine. The little ultralight, N41047, is just about the cheapest machine flying today—you can hardly fill the gas tank of a private plane for $49.95! Gunderson made a few changes in the original plane for safety's sake but basically it turned out to be a reasonable replica of the World War I trainer. "Those 1919 plans were just too light to be safe even for short hops," he points out.

In place of five-eighth inch ash longerons specified in the plans, he substituted one-inch thick spruce. He increased the rudder area by 50 percent and instead of using motorcycle wheel spokes for turnbuckles he used standard aircraft turnbuckles. For a wing, he picked up the pieces of a wrecked Aeronca and chopped down the span from 36 feet to 24 feet. After hanging a 650 cc snowmobile engine up front, he had to add 30 pounds of ballast beneath it to keep the center of gravity where it belonged.

Gunderson added an extra diagonal steel tube brace to stabilize the engine installation and built in jury struts to strengthen the wing bracing. The wooden prop, scrounged from a 40 horsepower Continental, is geared to the engine with a V-belt drive to provide 3:1 reduction.

Generally speaking, Gunderson followed the *Popular Mechanics* plans except for abandoning the USA-3 wing curve. He did follow the plans by starting construction of the fuselage first, from longerons and spruce struts and braces tied by piano wire. The wood, according to directions, "should be free from knots, pitch pockets and wind shakes, and coated with two coats of good spar varnish."

The engine mount was built of sheet steel, 72 by 32 inches in size. For the main gear, Gunderson used a pair of old Jenny wheels from his Pietenpol, fitted with motorcycle tires.

Conventional stick-and-rudder controls were installed and a light weight metal seat was fitted behind them. The gas tank was hung on the top longeron and finally the whole thing got a paint job prior to ground testing in the manner of the original Penguin trainers.

Gunderson recalls that the 1919 article was quite specific about there being plenty of room to roam in. "For flying," it specified, "a field, with at least one mile of straightaway and half a mile wide, is desirable. The first step is for the student to learn the manage his engine and steer with his feet. The plane should be started off down the field with all the controls in neutral position. Any tendency of the plane to swing off its course should immediately be counteracted by use of the rudder. When the student can taxi about under perfect rudder control, he is ready for his first hop.

"He should start out with the elevators slightly deflected and run along with the tail well up, until maximum speed has been acquired. Then, gently pulling the stick back, lift the machine 2 to 3 feet off the ground and push the stick back to neutral position. The hop should be continued for 100 yards or so."

The article wisely cautioned Penguin pilots against trying to fly in circles until they gained sufficient experience. Then, and only then, should they attempt to climb to the dizzy height of from 50 to 100 feet "before attempting circles, and they must be wide, without perceptible banking."

In his first hops, Gunderson found that he couldn't get the tail down far enough on takeoff to get the angle of attack he wanted. So he went down the pasture wide open at a hot 30 miles an hour. Once in the air, the Penguin handled just fine, he recalls. "The propeller slipstream was broken up by the engine so that it was about like riding a motorbike at 60 mph. You can really see the ground you're flying over!"

Because of all the extra drag of the open framework Gunderson used power down the approach until just off the runway. Then he greased her on at a nice 25 mph, rolling to a stop in under 100 feet.

Gunderson has retired his Penguin to the Experimental Aircraft Association's Air Museum, where it reposes as a fine example of an historic ultralight aircraft that started an unknown number of military aviators on their way to the high sky.

Chapter 4
Hang Gliders

The Wright brothers had to go through an evolutionary period of designing and flying gliders before adding power. The ultralight movement has largely developed from careful design studies in the early 1970s to improve upon the popular Rogallo wing hang gliders and learn more about their aerodynamic properties before adding small powerplants.

In 1973, the British Department of Transport was uneasy over the explosive growth of hang gliders in that country. They appealed for technical information from Taras Kiceniuk, director of the Mount Palomar Observatory in Southern California and a veteran sailplane pilot. His son, Taras Jr., a 16-year-old undergraduate at the California Institute of Technology in Pasadena, already was in the forefront of the hang glider movement. Young Taras was making remarkable flights in his Icarus II, a swept-wing, tailless biplane that weighed only 55 pounds empty and could be built for around $100 in about 150 to 200 man-hours. Icarus II used simple, effective wingtip rudders for turns and was capable of tight, continuous spiral turns with considerable dihedral and body shift. Stressed for 3 G load factors, its glide angle was 8:1 and sink rate 3.5 feet per second. Taras already had set a world record duration flight of two hours 26 minutes for foot-launched gliders, with landing at takeoff altitude, and another world record of 1000 feet altitude gain above foot-launched takeoff point.

Director Kiceniuk replied to the British officials: "I'd really like to explore the subject in depth, but that would mean writing a book.

I'd love to write such a book, but running an observatory is a time consuming task." However, he did make a number of pertinent observations on the basic characteristics of early hang gliders. He noted that gliders seemed to be sorting out into three main branches.

These, he said, were narrow angle Rogallo "kites," fixed wing and taut membrane hang gliders, and "back-yard" experimental types of the non-airworthy variety. The Rogallos with wide angle leading edges, he warned, "are dangerous and unpredictable, with pitch instability."

Kiceniuk suggested that "standards will have to be quite different for the different kinds of craft. So far, all craft which use fore and aft weight shift for pitch control are potentially hazardous when towed aloft by mechanical means, although this problem might be solved."

Stability

He continued: "There would seem to be an important difference between those craft which have a strong directional dependence on stability (similar to metacentric stability for boats), as compared to those which have a basic velocity dependent response. All hang gliders are a combination of both—but the ones which derive stability from the extreme pendular disposition of the overpowering weight of the pilot would seem to be unsuited for true thermal soaring or for flights in turbulent air."

The same is probably true, he added, "for membranes which can flap or lose their aerodynamic characteristics under extreme changes in attitude or angle of attack."

Spiral instability problems, Kiceniuk said, "should, I think, be of two types—using conventional control surfaces with insufficient effectiveness to roll out of steep, low-speed turns, and dynamic spiral instability, where hanging the pilot far below the craft can cause it to 'wind up' because of the centrifugal reaction." The control system on the *Icari* seems to be particularly effective for this class of flight.

Most of the non-towing accidents in hang gliders, Kiceniuk said, "seem to be associated with attempts to soar in conditions where the slope wind speed conditions cause the pilot to be blown backward into the hill or into lee eddies downwind of the ridge. Rogallo kites suffer due to relatively poor glide ratios. This tempts the pilot into seeking higher and higher wind speeds to permit soaring, rather than seeking steeper slopes. In light winds, a good

Rogallo probably affords the best and safest way to get down from a mountain—and that includes driving by automobile!"

Taras Kiceniuk Jr. went even deeper into the aerodynamics of his Icarus II flying wing hang glider that same year. In a seminar at Northrop Institute of Technology, he explained in detail why he considered the tail unnecessary and possibly harmful to the craft's flight and stability characteristics.

Longitudinal stability, he explained, can be achieved in any of several ways, "but in the end it always comes back to the same thing. The effective center of pressure and the associated lift vector acting on an airplane must move rearward if the craft noses up and forward if it noses down. The aeronautical pioneers understood this without knowing about moment coefficients, aerodynamic center, metacentric parabolas and so forth. For this reason, they performed their wind tunnel tests on complete wing-tail combinations to evaluate the suitability of various designs.

"Out of all this came the realization that the rear surface must, in effect, be at a lower angle of attack than the forward one. For a conventional wing-in-front, tail-in-back arrangement, this means that the tail is at a negative incidence angle with respect to the wing. For canards, the 'little wing' up front must be at a positive angle with respect to the main wing. For unswept flying wings, the designer makes use of reflex. He turns the trailing edge upward. For swept flying wings, the tips—being farther aft—are twisted down or washed-out, with respect to the center portion. Since this is what is needed to obtain a favorable span lift distribution and good stall recovery properties with a constant chord wing, we have killed several birds with one stone."

The Elevator

Discussing the function of the elevator (or elevon) Taras continued: "It functions to produce nose up or nose down moments on the aircraft, changing the angle of attack of the main airfoil—and with it trim the craft. With a conventional tail-in-back configuration, the elevator is so far behind the wing that the moment of change is quite large for small lift on the tail. The fact that the tail was producing a slight downward or upward force is of no consequence.

"In a flying wing, the situation is different. Raising the elevons produces the desired moment but at a cost. The elevon has, in effect, changed the camber—and with it the lift of the wing—so that the change in lift is less than the desired amount. When the elevon is deflected upward, the wing pitches up, giving more lift. At the same time the up-elevons also act like up-flaps, reducing the lift on the wing! The two effects counteract each other and make pitch control difficult. Both takeoffs and landings are tricky with this configuration.

"So what's the solution? How about center of gravity shift? If the angle of attack of the wing is increased by moving the pilot's body weight aft, the increase in lift is exactly the same as that produced by a conventional craft with a tail, for equal aspect ratio and angle of attack change. Interestingly enough, this is the longitudinal control system used by soaring birds. Jack Lambie has pointed out that a bird does not use his tail to change angle of attack, but rather to control his trim after he moves his wings forward or backward with respect to his body weight."

Lateral Control

Taras next discussed the problem of achieving lateral control and pointed out that, "given the intrinsically stable airfoil and sweep, one needs only dihedral angle and rudders attached to the swept wing tips to produce a completely flyable and steerable machine. Notice, I don't say anything about three-axis control. We must decide whether independent roll and yaw are important or even desirable. It has long been recognized that rudder paddles were added to most airplanes to make up for the shortcomings of their designers. As of this time, both Icarus II and Icarus V have demonstrated continuous spiral turns at angles of bank in excess of 55 degrees with no signs of instability.

"The other possibilities for lateral control are ailerons and spoilers. Ailerons have quick response, but they are tricky to build and actuate effectively. And they introduce problems of adverse yaw. Spoilers are simpler, but they are basically an inefficient, energy-wasting system. Considering the problems of these devices and the remarkable success I have had with the first Icarus, I have retained the simple, individually controlled wingtip rudders."

One drawback to the wingtip rudder system, Taras admitted, was that "roll control is slow for small corrections because the aircraft must yaw before it will roll. The moment of inertia is not negligible about the yaw axis. This is especially true for the 32-foot span Icarus V (Fig. 4-1). Still, the maneuverability of this craft actually exceeds that of most conventional sailplanes and gliders."

Aircraft employing wingtip rudders tend to be very stable, Taras says, because the effects of sweep and dihedral come into play. "In fact, they possess such strong static recovery characteristics that in a sense the pilot just goes along for the ride. Any disturbance from straight ahead flying automatically results in restoration of straight ahead flight. In a normal turn, neutralizing the differentially controlled wingtip rudders means a spontaneous return to normal flight."

The Airfoil

In selecting an airfoil for the Icarus II wing, Taras went to a modified Eiffel section chosen for its low moment. The one that is used on Icarus V is more sophisticated. He explains, that "after studying the results of Stratford, Lissaman, Liebeck, and Wortmann, an airfoil was 'eyeballed' which would have low moment and still possess high efficiency at high lift coefficients."

Low moment is achieved by employing reflex in the camber line, Taras explains. High efficiency is attained by generating a shape with a generous leading edge radius and a rather abrupt thickening, followed by a smooth, gentle, "recovery" section. "Several such sections were analyzed," says Taras, "thanks to the generosity and ingenuity of Dr. Peter Lissaman, using the Hewlett Packard computer at Dr. Paul Macready's AeroVironment, Inc. firm in Pasadena. The resulting pressure distribution and performance characteristics seemed ideal for a craft of this type and flying in this regime."

While higher maximum lift coefficients and lower stalling speeds can be realized with highly cambered, high moment airfoils, says Taras, recent experience has shown that takeoff and landing speeds are well within the capabilities of most pilots without having to rely on ultra-light lift sections. "Indeed," he says, "most pilots would trade extremely low stalling speeds for high speed performance. Better high speed performance can be expected from sections without large undercamber.

An Ultralight Glider

The final design of the Icarus II and Icarus V was an almost inevitable consequence of several individual design considerations, each interlocking with, and crying out for the other. The result is a stable, pilot-launched ultralight glider with a glide ratio of about eight or ten to one, a large usable range of flying speeds and excellent controllability."

Success of the Icarus II biplane design is evident. It has become the most popular configuration for adaptation to powered ultralight flying, as pioneered by John Moody of Milwaukee, Wis-

Fig. 4-1. Taras Kiceniuk Jr. designed the lovely Icarus V hang glider with wingtip "rudder-vons."

consin in the Easy Riser version. Since many Icarus V monoplane hang gliders are being converted to power, let's look in on its early development as described by Taras Kiceniuk Sr.

Icarus V is a swept, constant chord, flying wing monoplane of 32-foot span and 5-foot chord. It has a wing area of 160 square feet and an aspect ratio of 6.4. Construction is of aluminum tubing, cable (three-thirty-second of an inch) braced and fabric covered. Its efficient, computer analyzed, high-lift airfoil (TK 7315) has a foam-sheet leading edge formed over aluminum ribs. It was designed with an ultimate load limit in excess of 6 G's for a 200-pound pilot. Empty weight is only 65 pounds. Construction time is about 200 man-hours. It disassembles at the centerline in 10 or 15 minutes for transportation in a cartop box. Assembly time is only a little longer.

Control is by the same system used in the earlier Icarus II—pitch control by body weight shift on parallel bars. The pilot sits on a swing seat. Lateral control is through individually controlled tip rudders. Deflecting a rudder causes the craft to yaw and the combined sweep and dihedral produce strong roll response. Thanks to the 20-degree sweep and the 7-degree geometric twist for washout, the Icarus V cannot be stalled accidentally. And pitch stability is astounding!

With a glide ratio of around 10:1, says Kiceniuk, landing in a small area could present something of a problem. To meet this, both tip rudders can be deflected simultaneously to serve as dive brakes. The resulting drag provides effective glide path control.

"Unlike most ultralights," Kiceniuk says, "especially those with single-surface sections, Icarus V can be flown fast without excessive loss of glide ratio, a must for cross-country flying and for penetrating out of tight spots in strong winds near the crests of hills. It is estimated that the V can be flown at 45 miles an hour when nosed down and will still have a better glide angle than a standard Rogallo wing flying at its best L/D. At the slow end, Icarus V can fly at 16 miles an hour with sinking speeds around one meter per second."

Kiceniuk explains that Icarus III and Icarus IV were designs that never got off the drawing board when it was decided to concentrate on Icarus V, which was introduced at the third Montgomery Meet in 1973.

Chapter 5
Ultralight Engines

Selecting the proper powerplant for an ultralight aircraft is largely a matter of using Hobson's Choice. In the 17th Century, a liveryman named Thomas Hobson had a simple answer to customers wanting to pick the best horse in the stable—"Take the one nearest the door, sir!"

Today homebuilders frequently resort to the same logic in selecting an engine—"Take the one that is available, sir!"

Long before the Wright Brothers built their engine for their Kitty Hawk flyer, aerial experimenters were trying steam, electricity and all sorts of other power sources to achieve light weight and adequate horsepower. The power-to-weight ratio is what we call it today. Motorcycle engines were in big favor in the first decade of this century, but as airplanes grew bigger and faster their engines grew heavier and more powerful.

Now that the trend has reversed itself and the era of ultralights has arrived, the idea that big is beautiful is passe. VW engines have literally been cut in half to make lightweight, two-cylinder powerplants that really work. A few motorcycle engines have been modified to fly. Outboard motors, snowmobile and chainsaw engines have been widely accepted. And a few hardy souls have actually developed original engines for ultralights.

A Go-kart Powerplant

John Moody's powered Easy Riser (Fig. 5-2) of the mid-1970s pointed the way with a modified McCulloch 101 engine that has

Fig. 5-1. The Soarmaster C5A Rogallo-type ultralight uses a 10-horsepower Chrysler powerplant.

become perhaps the most widely used powerplant in the field. The McCulloch go-kart powerplants run from 100 to 125cc displacement, with a power range of from 10 to 18. These are high-rpm engines with crankshafts that were not designed to take the loads imposed by the demands of flying such as full-power on takeoff and engine-idle letdowns. Running at high rpm also imposes a resultant shock-wave damage. The bigger "Macs" frequently are powered with Methanol and Nitro methane fuels for bursts of higher power—in the 24 horsepower range—but the higher torque cuts engine life severely.

Snowmobile Engines

Scroungers have found that some snowmobile engines work well. Examples are the Chaparral 242, which weighs about 32 pounds and delivers some 20 horsepower, or the JLO 397, a 45-pounder that can put out 36 horsepower. These single-cylinder air cooled engines are fairly reliable and operate at direct drive speeds. Reduction units operate at a noisy 6500 rpm.

Another choice is the Chrysler 820 industrial engine with 8 to 10 horsepower. It is a two-cycle powerplant that swings a bigger prop via a reduction unit, as in the Soarmaster design. This power train runs up to 35 pounds and the engine can be uprated to 14 horsepower, which unfortunately cuts its reliability (Fig. 5-1).

A New Design

John Chotia, designer of the popular Weedhopper ultralight, decided the way to go was to design and build a new engine specifically for ultralights—Weedhoppers and others. Says Chotia: "When we started the Weedhopper program, we had some 250 snowmobile engines lined up, but they were quickly snapped up. The real delay in our engine program began when we were unable to achieve the 50 to 55 percent prop efficiency we wanted, due to the lower tip velocity. Our Chotia 460 propeller tip velocity was only .67 Mach, but our efficiency was only about 37 percent compared to the snowmobile engine-propeller tip velocity of .95 Mach at peak power and a low propeller efficiency of about 25 percent. As a result, our performance of 18.3 horsepower was too low for what we wanted.

"The culprit here," Chotia went on, "is the low air mass passing through the propeller disc. The Weedhopper's 42-inch prop at 27 mph airspeed puts three times as much energy into each pound of air as a 72-inch propeller swung by an engine of 100 horsepower flying at 100 mph. The result is that the air virtually slips away. The prop wash must move at from 65 to 70 mph to provide sufficient thrust, almost like a jet blast, as opposed to the 100-horsepower, 100-mph combination, where the propeller acts like a rotary wing."

As a solution, Chotia saw that he needed more power. He designed it into the engine and the result was the production Chotia 460 engine which delivers 25 horsepower at 3600 rpm and weighs only 32 pounds. The engine sells for $700 (less propeller) from Weedhopper of Utah Inc., P O Box 2253, Ogden, UT 84404.

The Gemini System

Another interesting power system is Ed Sweeney's Gemini Twin Thrust System (Fig. 5-3). Says Sweeney: "To power a hang

Fig. 5-2. The starter rope gets a yank from an Easy Riser pilot. He can stop and restart the engine in flight.

Fig. 5-3. Ed Sweeney's Gemini twin-engine powerpack includes special guard fenders for propellers.

glider, some unique problems must be solved. The Gemini System is a twin-thrust arrangement which locates the thrust line as close to the center of mass as possible and the weight of the system is on the center of gravity. This eliminates the adverse handling effects associated with thrust lines that vary as the center of gravity is moved by the pilot's shifting body weight.

"Also, as a safety feature, the pilot has no obstruction either in front or behind. Since the engines are near the center of the machine, they pose very little or no asymmetric thrust effects in single-engine flight."

Designed for all types of hang gliders, the Gemini system mounts on the basic triangular pilot cage and the pilot is isolated from the blades by shields and the A-frame. Each engine has its own

Fig. 5-4. The Soarmaster powerpack is mounted on a Cirrus 5 Flexwing. This keeps the propeller blades at the rear and away from the pilot's feet.

Fig. 5-5. The Soarmaster PP-106 powerpack has a two-stroke, 10-horsepower Chrysler engine.

throttle, with the rpm fully controllable. The modified engines built by Gemini International Inc., 655 Juniper Road, Reno, NV 89509, use electronic ignition, recoil starters, piston port induction (no reed valves), a pressure pump carburetor and weigh only 8.5 pounds each.

The standard Gemini Twin Thrust System delivers 85 pounds static thrust (60 pounds dynamic thrust) and weighs 28 pounds total. Each engine delivers 6 horsepower. Propellers are 26:25, fuel capacity is .6 U.S. gallon and duration is 24 minutes at full power. Noise level is 93 dB at 15 feet.

The Soarmaster

The Soarmaster Powerpack-106 is another popular installation for flex wing hang gliders of the Rogallo variety (Fig. 5-5). Instead of jumping off a mountain, today's PHG glider rides take off from the flatland and climb into thermal country. With two quarts of gas you can stay up for hours, shutting down the engine and restarting when the lift goes away.

Installed in 3 minutes on a Cirrus V hang glider (Fig. 5-4), the Soarmaster gives it a cruise speed of 32 mph, maximum velocity of 40 mph and a 325-fpm climb to a ceiling of 7000 MSL. The Soarmaster Powerpack uses the 2-stroke Chrysler 10-horsepower engine (Fig. 5-5) that turns up at 8000 rpm. A chain-drive reduction unit is

Fig. 5-6. Larry Mauro's Easy Riser has some 500 solar cells on top of the wing to power a battery that runs an electric motor.

linked to a 4130 chrome moly steel tube drive shaft, driving a 42:19 propeller at the rear. Total weight is 27 pounds. For information contact Soarmaster Inc., P O Box 4207, Scottsdale, AZ 85258.

The Powerhawk

A top-of-the-line system is the CGS Powerhawk designed by a veteran glider rider, Chuck Slusarczyk, president of CGS Aviation Inc., 4252 Pearl Road, Cleveland, OH 44109. As installed on the Easy Riser, CGS Powerhawk retails at $875 and includes a 10-horsepower West Bend and Chrysler engine, 42-inch propeller, pulleys, belts, reduction mount with bearing, muffler, engine mount assembly, throttle, kill switch, fuel tank, fuel line, hookup wire and all other required hardware.

CGS Aviation also has a conversion pack for the Mac 101 engine, retailing for $625. It delivers up to 85 pounds static thrust. CGS has developed the first aircraft style engine mount and the powerplant system is currently used on such top PHGs as the Mitchell B-10 flying wing, the Birdman TL-1A and Volmer Jensen's VJ-23 Swingwing and VJ-24E Sun Fun.

In a time of high fuel costs, Larry Mauro, president of Ultralight Flying Machines in Santa Clara, CA, amazed everyone recently by using arrays of solar cells to capture sunlight to charge a battery to run an electric motor. He flew it successfully on April 29, 1979, at Flabob Airport in Southern California (Fig. 5-6).

Says Mauro: "There is no energy crisis—there never was one and there never will be! We needn't look underground for sources of power. All the power this earth can use is outward and upward. The sun is virtually a limitless source of power and just a wee fraction will enable mankind to accomplish purposes as yet undreamed!"

Chapter 6
Powered Hang Gliders

When a veteran sailplane driver turned to the design of an ultralight powered aircraft, it was almost a foregone conclusion that it must have high performance in thermalling and ridge-soaring as well as ruggedness, quick assembly time and a good safety margin.

The Sun Fun

Such were the parameters considered by Volmer Jensen when he elected to join the mushrooming ultralight movement recently. With the assistance of another veteran designer, Irv Culver, Jensen produced the excellent VJ-24E Sun Fun powered hang glider, with a rigid wing and conventional aircraft controls (Fig. 6-1). See Figs. 6-2, 6-3 and 6-4 for other examples of powered hang gliders.

Outstanding because of its gracefulness, Sun Fun attracts admiration whenever and wherever it appears at powered hang glider meets. An example was the second annual Diamond PHG Meet at Perris Valley Airport in Southern California in the spring of 1979. While standard Rogallo gliders, Quicksilvers and Easy Risers sat ground-bound in gusty afternoon winds, Vol Jensen was up, up and away doing his thing high overhead while other pilots watched in awe.

To grasp the significance of Sun Fun's performance, you have to understand the man who conceived it. To understand Vol Jensen, you must know something of his background in flying. Back in 1931, when Charley and Anne Lindbergh took up soaring along the ridges

Fig. 6-1. Volmer Jensen's rigid-wing VJ-24E Sun Fun is an excellent ultralight.

near Gorman, at the south end of California's San Joaquin Valley, Jensen already was an accomplished sailplane driver.

By 1941 Jensen had produced his tenth glider, the J-10 two-seater. Abbott and Costello borrowed the J-10 for a publicity stunt to sell war bonds and the War Department drafted it to train glider pilots. Right after World War II, Jensen came up with an unusual side-by-side two-place sport plane with a pylon-mounted pusher engine. It was called the VJ-21 Jaybird and I had fun flying it down to an air show at Palm Springs for Vol.

Next came his popular amphibian two-seater, the VJ-22 Sportsman, which he used to explore hidden coves on scuba diving adventures in the Sea of Cortez.

With the rebirth of the hang glider movement in the early 1970s, Jensen studied the crude Rogallo kites built of bamboo and

Fig. 6-2. Many Icarus V hang gliders have been converted to power. This one belongs to Ted Ancona.

Fig. 6-3. Taras Kiceniuk's Icarus II hang glider became the Easy Riser when an engine was added.

polyethelene sails and decided there was room to join the foot-launch gang with something that would have better flight characteristics than the prevalent ground skimmers. He looked up Irv Culver, a friend and retired Lockheed aerodynamicist. Together they sketched plans for a superlight, boom-tailed hang glider with a 32-foot rigid plywood wing. Culver contributed the high-lift airfoil and ran the stress analysis on the whole structure to insure that it was safe. The VJ-23, as it was designated, weighed 100 pounds and differed from all previous hang gliders since it was fully controllable. Culver added a joystick to dispense with the need for rudder pedals to coordinate rudder and aileron control with the single stick.

On its initial trials, the VJ-23 performed far better than Jensen had anticipated. Launched from a 75-foot hillside with a broad 3-to-1 slope, Jensen ridge-soared for well over five minutes before deciding to land. At the Playa Del Rey hang glider site near Los Angeles, Jensen and the VJ-23 joined the parade of gulls working the ridge for 42 minutes. It was the first public demonstration of duration flying in a hang glider.

The VJ-23 was an immediate success. When *Mechanix Illustrated* published plans for it, the machine became world famous. In England, one homebuilder won a cash prize for outdistancing all other hang gliders in a launch from a 20-foot high platform.

The VJ-24

In 1974, Jensen followed the VJ-23 with a new model. The VJ-24 was made of all-metal construction and pop-riveted together in a fashion that required only 200 man-hours to construct. Wings and tail were snap-locked on with pip pins rather than nuts and bolts, making it easily and quickly assembled for flight or disassembled for transport.

To launch the VJ-24, Jensen, who weighs 135 pounds, does not have to run forward. A few quick steps into an 8 to 10 mph wind gets him airborne. In a 15-mph headwind, the machine lifts its own weight at a standstill. Landings are made at jogging speed. With practice, toughdowns can be birdlike—the way a sparrow settles onto a branch.

In flight, the VJ-24 is responsive to the slightest control pressures and easily maneuvered for thermalling or ridge soaring. In one flight along the seashore cliffs of Torrance, California, Jensen stayed aloft for 90 minutes. He gained 200 feet of altitude on each run up and down the cliffs. On one pass he silently buzzed a cocktail party at a hillside home. He reported, "the guests all leaped to their feet, glasses raised high, and cheered me on!"

If the VJ-24 Sun Fun had unlocked the secret of the birds, there was still a new dimension of flight to be added. It was well and good to be able to soar along grassy ridges where gentle slope winds blow, to take off with only a couple of steps and a hop and to land at zero velocity with a simple knee bend.

But the era of powered hang gliders had arrived and Jensen instinctively saw that by adding a light power plant and propeller, a VJ-24E Sun Fun PHG could open a whole new realm of flight. Powered with a McCulloch 101 go-kart engine and a tiny pusher

Fig. 6-4. Volmer Jensen was a pioneer in the powered hang glider movement with his graceful Swing Wing.

prop, the VJ-24E can outperform the best powered hang gliders by launching from level ground at the foot of a slope and flying uphill to soar along its crest, instead of the other way around.

The result of 14 months research and development, is the outgrowth of performance studies utilizing three different powerplants, expansion exhaust chambers, eight muffler designs, two carburetor manifolds, four rubber shock mounts and eight propeller configurations. Brochures describing the engine installation on a stock VJ-24 are available from Volmer Jensen, PO Box 5222, Glendale, CA 91201.

Curious as to who were buying VJ-24E Sun Fun plans and kits, Jensen surveyed the builders and discovered that "where the average hang glider pilot is young, the Sun Fun apparently appeals to an older generation—lawyers, doctors, businessmen attracted to the PHG movement who want something easy to build and easy to fly safely with a little power added."

Sun Funs are flown regularly from established hang glider sites. Some are known to be operating from designated airports, FAA-licensed and operated by at least student pilots holding valid airman certificates.

However, Jensen advises against flying ultralights in the vicinity of airports even when legally licensed to do so. "For obvious reasons," he says, "a lightly-loaded hang glider is likely to become unmanageable in the wake of even a small, low-powered aircraft such as a J-3 Piper Cub. Most hang gliders are difficult to see head-on or from behind in flight. Their low speed makes them highly vulnerable in an airport traffic area."

This is particularly true of PHGs and unpowered hang gliders being flown from ridges near the approach and departure lanes of nearby airports. An example is the Sylmar site at the north side of San Fernando Valley, CA. It is close by four major airports—Van Nuys, Burbank, San Fernando and Whiteman. Powerplane pilots already have reported several near-misses.

Out in the open country, however, Sun Fun is a joy to behold and fly. One spring day I watched Vol Jensen demonstrate his Sun Fun from a freshly-plowed field near Thousand Oaks, CA. It is a rural area where ocean breezes sweep inland and up-slope over a grassy, 500-foot hillside. A tractor stopped nearby and its operator advised that this would be our last opportunity to fly there for some time since he was preparing to plant a crop of lima beans. Jensen nodded. It was the old story, hunting for virgin hillsides facing into the prevailing wind, where one can escape the mundane worldly cares and enjoy briefly the closest thing to bird flight.

"Up there on top of the hill," Jensen pointed. "That's where we used to launch hang gliders. But now I can launch from down here with a couple of steps and ride the slope winds up to the top."

No sooner said than done.

As I watched, Jensen and a friend removed the Sun Fun from its trailer bed. Within 10 minutes, the two of them had it all set up and ready for flight. The job was made easier by the special lock pins that held the wings, tail and struts firmly in place. Not even a screwdriver or a pair of pliers was required.

Jensen stuck a portable wind sock mast into the ground and consulted his little vest pocket wind meter. The tiny ball gently bounced up and down close to the 15-knot velocity mark. Just right. Next he poured a quart of gasoline into a plastic tank, checked everything in a careful preflight final inspection, stepped into position between the sides of the aluminum tubing framework of the hanger and tucked the spars under his armpits.

Sun Fun weighs 110 pounds empty. With its engine installed it weighs roughly 150 pounds. Facing into the gentle wind, Jensen lifted it easily with the help of mother nature. The engine started with a quick pull on the starter cord, positioned for aerial restarts. Sun Fun trembled expectantly, anxious to be off and flying. Jensen took two quick steps forward, knees bent, then tucked up his feet for a very quick takeoff. He slid agilely into the sling seat, manipulated the control levers expertly and buzzed off.

As I watched in amazement, he moved slowly over to the foot of the hill at perhaps 20 miles an hour. He banked over like a hawk sniffing for the slope wind. He found it quickly, sweeping through a graceful 360, then began soaring higher in lovely figure eight turns until he topped the ridge. For long minutes he ridge-soared with the engine throttled back and paced by a pair of curious gulls. He could have stopped the engine and still remained aloft for an easy hour of powerless soaring flight in the afternoon onshore breeze. Instead he elected to sweep back down the hill to demonstrate a landing for the benefit of my camera.

From where I stood, partway upslope, I watched the Sun Fun bend around into the wind and gently swoop down with birdlike effect. Vol tilted the wing upward a bit and came in for a zero-speed touchdown. His knees were flexed to absorb the energy of the landing.

Another time he chose to land atop the hill, shut the engine down and trundle the machine by the tail into the slope wind for takeoff. With four or five quick steps he was off again, this time into an eight-knot headwind. It demonstrated the advantages of Sun Fun flight over that of conventional hang gliders, which must be disassembled at the termination of each flight and trucked back uphill for the next joyride.

Irv Culver, who weighs 200 pounds, has flown Sun Fun a number of times. His skill as a sailplane pilot and a conventional airplane pilot make up for the higher gross weight. As in any aircraft, flying skill makes all the difference between good performance and poor performance. Jensen recommends that beginners take a couple of hours dual instruction in a light aircraft with stick control, such as a Piper Cub or an Interstate Cadet, to get the feel for Sun Fun piloting. Of course, this is not a mandatory procedure. "Any sensible, healthy individual from 16 to 60 who builds a Sun Fun should be able to teach himself to fly it safely without difficulty," says the designer.

Design parameters for Sun Fun included easy construction and good performance plus inherent safety, including a design load factor of two and an ultimate load factor of three. Sun Fun is all metal except for the wing fabric, built mostly of aluminum tubing including the wing spars. It took Jensen only four hours each to assemble the rudder, stabilizer and elevator, less the covering.

"You just cut the tubing and pop-rivet it all together," he explains. "Total time required to complete Sun Fun runs around 200 hours, about half that needed to build the VJ-23 Swingwing."

The hanger structure is designed to insure safety of the pilot in the event of a belly landing. Wheels were added to the design simply to make it easier to transport. You can roll it where you want it instead of carrying it.

Where Sun Fun doesn't have the lovely tapered wing planform of the Swingwing, its higher aspect ratio compensates for the extra drag of external lift struts. And you can buy a kit option for Sun Fun that includes a protective device, not as a landing gear, to keep it in the category of a foot-launchable PHG and not a standard airplane.

Across the Channel

Sun Fun made international headlines in May, 1978, when one built by British draftsman David Cook from Leiston, Suffolk, was flown easily across the English Channel from the Kent coast near historic Walmer Castle, 25 miles to the coast of France. Reported the Daily Mail:

"David strapped on a cherry red helmet, donned a wet suit, gave his wife, Cathy, a kiss and ran like the clappers for France while a competitor, Gerry Breen, scanned the horizon for his escort boat. An hour and a quarter later, with just enough fuel left to wet a postage stamp, the 37-year-old father of two landed beside a party of picnicking Germans on holiday in the resort of Bleriot Plage. It was an appropriate landing strip—it was near where French aviator Louis Bleriot took off in 1909 for the first air crossing of the Channel."

A year later another odd aircraft would complete the challenging Channel crossing. Dr. Paul MacCready's man-powered *Gossamer Albatross* was pedaled furiously by veteran bike racer Bryan Allen. Man has yet to cross the Chann l in an ornithopter, however.

At this writing, there were several hundred sets of Sun Fun PHG plans sold at $55 each. They comprise five blueprints totaling 36 square feet, 24 photos and a full size rib layout. A $2 brochure also is available for the VJ-24E Sun Fun engine installation from Volmer Jensen, PO Box 5222, Glendale, CA. 91201. See Table 6-1.

Quicksilver

In 1972, wher the hang glider movement was just rolling, a sharp little high winger began showing its stuff at the local glider slopes. Quicksilver (Fig. 6-5) called a classic in its own time, had the traditional aircraft look and was quite stable in flight. It did not have a tendency to tuck under in a dive as Rogallo flexwings sometimes did.

Fig. 6-5. The Mitchell Wing B-10 is designed with folding wings for easy transport.

Table 6-1. VJ-24E Sun Fun.

Dimensions	
Length	18′
Height	6′
Span	36½′
Wing Area	163 sq/ft.
Weight	
Empty	110 lb.
Useful Load	200 lb.
Gross Weight	310 lb.
Performance	
Crusing Speed	20 mph
Stall Speed	15 mph
Construction	
All Metal	
Fabric Covered	
Controls	
Ailerons, Elevators, Rudder, Throttle	
Glide Angle	
9 to 1	

Quicksilver was the first monoplane hang glider to be made available complete or in kit form by Eipper- formance Inc., which remains a leader in the field today. Eipper-formance Inc. has been a major manufacturer of microlight aircraft since 1972. At all major homebuilt aircraft meets, their top-of-the-line Quicksilver stands out for its good looks as well as excellent performance.

Since its introduction a decade ago, thousands of Quicksilvers have logged thousands of hours the world over, while maintaining an excellent safety record. Quicksilvers were the first rigid wings employing simple bolt-together construction. Its materials came directly from the flex-wing field. Its design was so basic that, the company says, it has changed little since 1973 except for improved hardware and small detail modifications.

Quicksilver was the first hang glider to make an altitude gain of a mile above takeoff point. It also had climbed to 15,000 feet MSL and covered more than 20 miles on cross-country flights. The design was particularly attractive to pilots who felt comfortable with its conventional configuration.

Early on, Quicksilver established its reputation for pitch stability, largely attributable to its conventional empennage design. When the era of powered hang gliders arrived, Eipper began an exhaustive series of tests of power plants, propellers, landing gear systems and various airframe mods. The result was what Eipper considers the best combination of power-on and power-off handling characteristics.

Says Eipper: "Our philosophy in motorized design has been to reduce flight to its essence. We have found that through continual simplification and refinement of our designs, safety, reliability and cost reduction can be greatly enhanced."

Ground handling the Model M Quicksilver, says Eipper, is relatively simple. Prop wash over the rudder in combination with the lightly loaded nosewheel make rudder inputs very effective upon directional control. At rotation speed (16-20 mph) the pilot simply pushes his weight back to lift the nosewheel and commence the climb-out. All pitch control is accomplished by means of pilot weight shift, eliminating weight and balance problems that might occur with the pilot accounting for half the gross weight.

Lateral control is achieved simply by the pilot shifting his weight in the direction he wants to go. The pilot's harness is connected through control lines to the rudder. A weight shift to the left turns the rudder left, yawing and rolling the aircraft into the turn. Quicksilver's generous dihedral gives a light and quick roll response and also provides a high roll and spiral stability.

The main wing is rectangular in planform and is single surface, keeping the wing as light and simple in construction as possible. Wingtip washout in combination with the wing's taperless planform, says Eipper, makes the stall extremely gentle. There is no tendency to drop a wing or enter a spin.

In flight, the engine could be shut down for soaring or climbing in lift and then restarted by pulling on the starter cord. For landing, power is reduced and the approach speed adjusted to between 20 and 25 mph. Landing in a headwind of from 10 to 12 mph, touchdown velocity is little more than fast walk. This allows for extremely short-field landings.

Quicksilvers make good transition trainers for pilots planning to move on to conventional aircraft. Owners of the earlier Model C

Quicksilver hang gliders can now purchase retrofit powerpack and landing gear kits to convert to the Model M powered hang glider configuration.

The complete Quicksilver M, ready to fly and factory flight tested, comes with detachable landing gear for foot or wheel landing and launching and comes complete with heavy duty nylon storage covers for $3495. The Model M kit required only from 20 to 30 hours construction time using basic tools. It includes assembly instructions and all required materials, including a supine pilot harness. Also included are all tube cutting, drilling, anodizing, tube bending where required, cable swaging, and completely finished wing and tail surface covering.

The engine, a Chrysler 82026 two-cycle, dual carburetor with 137cc displacement and 13 horsepower output, has a thrust rating of 110 pounds. A 1.7-gallon fuel tank is provided. Fuel consumption runs from one to two gph, depending on power setting. The kit price is $2995.

Quicksilver M's wingspan is 32 feet, wing area is 160 square feet, aspect ratio is 6.4, empty weight is 130 pounds and the pilot weight range is from 120 to 220 pounds. Performance with a 160-pound pilot, no wind at sea level, is: 20-25 mph cruise, 35 mph Vmax, 17 mph stall, takeoff roll 100 feet, landing roll 50 feet and rate of climb 350 fpm.

Eipper-formance Inc. is located at 1070 Linda Vista Drive, San Marcos, CA 92069.

A Foot-Launched Air Cycle

Five years of research and development work went into one of the later entries in the ultralight arena. Ken Striplin's 156 pound, chainsaw-engined, tricycle-geared flying wing pusher is called FLAC—for Foot-Launched Air Cycle. First flight took place October 25, 1978, at El Mirage Dry Lake north of Los Angeles (Figs. 6-6, 6-7, 6-8, 6-9). The pilot was Ken's teenage son, Paul. Results encouraged the Striplins to go ahead and put the craft into a production run. Their goal was one a day.

Paul's initial test hop proved the craft was a stable, responsive, easy to fly powered hang glider with three-axis controls. He circled the dry lake bed for half an hour at speeds up to 45 mph and at one point deliberately shut down the McCulloch 101 engine, glided a bit, then restarted with two pulls on the starter rope.

It appeared to Paul that the center of gravity was a bit far forward, resulting in a nose-wheel-first landing that bounced him

Fig. 6-6. Test pilot Paul Striplin, 19, flies the FLAC ultralight from El Mirage Dry Lake in California.

back into the sky for a go-around. Next landing was perfect. Back in their shop in Lancaster, they moved the wing forward three inches and then Paul took her up again. This time it flew fine hands off. After liftoff at an estimated 21 mph, FLAC climbed at 25 mph for some stall tests. Paul found the nose eased down at 20 mph IAS. At cruising speeds from 25-30 mph, FLAC was easily controllable. When it wouldn't come down due to a too fast engine idle, Paul killed the engine and made a nice deadstick landing. The engine was readjusted and the wheels moved back a bit for better three-pointers.

Other modifications followed, particularly in the power reduction unit. An enclosed chain and oil clutch lashup was developed that

Fig. 6-7. FLAC has wingtip rudders, elevons and tricycle gear.

Fig. 6-8. A modified FLAC with extended nose gear for faster takeoffs.

is similar to the Soarmaster power pack. Soft engine mounts were used to cut down vibration. An expansion chamber muffler pointed up and cut back the noise level appreciably. Later, Ken installed a coupled twin-engine powerplant driving a single prop.

Although FLAC has a tricycle landing gear, there are clamshell doors that open to permit foot-launching. This makes it a non-airplane by FAA definition, says Striplin. Recently a longer-legged, retractable nosewheel gear was used to permit the craft to take off at a slightly higher angle of attack on the ground run.

At one point, the Striplins decided there was not enough elevon control at low airspeeds. They lowered the control surfaces a bit below the trailing edge to keep them in harder air on approaches. This was similar to the Mitchell Wing design.

To further improve the control response, they decided to reduce the area of the wingtip rudder plates by roughly 50 percent. The extra surface was found to be unnecessary and produced drag. The FLAC wing sweeps back gracefully with a constant chord. A "special" German airfoil, says Striplin, develops high lift with low drag and pitch.

Controls are three-axis and operated by the pilot just as in a conventional aircraft. Pitch and roll is achieved by a side stick linked through a mixer to operate the control surfaces together as elevators or differentially as ailerons. Twin tip rudders are controlled by foot pedals linked with cables to activate them independently or together as air brakes. The rudders also serve as span airflow control fences. Striplin says they increase lift further and reduce tip drag.

The FLAC cockpit has been modified. It is made of fiberglass fitted with a large windscreen and side openings to give 180-degree visibility. Recently, the side openings were covered with Mylar for better streamlining. The pilot sits in a hammock seat so that he can swing his legs down for foot-launch. The main wheels are enclosed in wheel pants, while the nose wheel retracts after takeoff.

Earlier, the Striplins had built two craft with conventional tails and three flying wings, plus numerous RC models. More recently, they experimented with the development of an ultralight based on the flying wing sailplane design of Witold Kasper and rights were obtained to construct a prototype. However, it was abandoned as being too unstable.

By February 1978, the Striplins had built and flown half a dozen RC models of one-third scale, seeking a design to freeze with good pitch stability and directional control. In deliberately tumbling the RC models, they determined that recovery was immediate and that adding extra power was a no-no, without a major redesign of the flying wing.

Their fourth design confirmed their RC work. By auto-towing that one, they decided to go ahead with a fifth design—FLAC. The prototype was completed in one month.

FLAC's wing is easily removable by taking out six bolts for road transport. Plans are shaping up to add skis to turn the air cycle into a snow cycle or floats to make a water cycle.

Kits are available from the Striplin Aircraft Corp., P O Box 2001, Lancaster, CA 93534, at $1895 for the economy model or $2895 for the "quick flight" version. Engine is not included, but Striplin recommends a Soarmaster powerplant. An information booklet is available for $5.

Fig. 6-9. Designer Ken Striplin folds the wings of his FLAC ultralight after an airshow at Chino, California.

Fig. 6-10. These two Fledgling Pterodactyls flew coast-to-coast in 1979.

Pterodactyl Fledgling

Big news at the Oshkosh '79 EAA Fly-In ultralight arena was the arrival of a pair of Californians who had flown all the way from Monterey in Pterodactyl Fledgings (Figs. 6-10, 6-11, 6-12). It was a barn storming tour that reached altitudes in excess of 14,000 feet crossing the Tetons and took just under 74 hours flying time. The intrepid pilots, Jack McCornack and Keith Nicely, stopped when and where they pleased. They landed in vacant lots or on golf courses and saw America roll by quite like no other airmen ever had. After a week at Oshkosh, they took off once more and continued

Fig. 6-11. A powered Fledgling at the Oshkosh EAA Fly-In.

Fig. 6-12. A Pterodactyl lands on bicycle wheels for use on rough strips.

east to Kitty Hawk, NC. They traveled that distance on an alternate fuel—grain alcohol.

A wild adventure for sure. But their arrival at Kitty Hawk was preceded by another Pterodactyl Fledgling, in which Jack Peterson completed a coast-to-coast jaunt from Los Angeles. Peterson's flight was reminiscent of the first west-to-east transcon flight by Bob Fowler, in a Wright biplane, in 1911.

To understand the beauty of these adventures, you have to understand that a Pterodactyl Fledgling is a sort of evolutionary derivative of an earlier maverick—Manta Products' early single-surface Fledge. The Fledge was developed from a design created in 1974 by Klaus Hill, Larry Hill, and Dick Cheney.

Fig. 6-13. The Mitchell Wing B-10 is designed with folding wings for easy transport.

Manta Products spent three years developing the Fledge. The goal was to combine the simplicity of the Rogallo wing with the better performance advantages of a rigid airfoil hang glider. The Fledge was designed for better penetration and speed for best minimum sink and LD by utilizing preformed aluminum ribs and control surfaces instead of weight shift. The stiff ribs meant you could shape and maintain the sail's form, camber and reflex without further adjustment or rigging. A lower surface was added to improve low-speed handling.

The result was called Fledge II, which not only offered better low-speed performance, but allowed faster turns, a better LD, higher Vmax and easier ground handling. The original airframe was unchanged and the first protogype, called the B Model, used an increased camber of seven and one-half percent at quarter-chord. A tip-mounted rudder and booster tip were added and the rudder surfaces raked back 24 degrees to provide a "semi-aileron" effect. The tip surfaces also served as end plates to reduce span flow and the associated vortex drag.

The Pterodactyl Fledgling was the next step. It was an ultralight motor glider based on the Manta Fledge II-B hang glider design, offered by Pterodactyl, of 847 Airport Road, Monterey, CA 93940. The Pterodactyl Fledge is fully collapsable for transport or storage. The kit is designed to meet the FAA requirement of 51 percent amateur built construction to qualify for an experimental license.

Jack McCormack explains that the "steps involved in constructing a Pterodactyl Fledgling are, airframe assembly, rigging, landing gear assembly, power unit assembly and wing cover installation. It is recommended that you allow a full day for rigging, but each of the other steps can be done in an evening or two.

"Though we can't tell you how long it will take to build yours, at Manta, a dedicated crew of two builds three Fledgling hang gliders a day. On the other hand, if you work on it after work and before dinner, figure a month. No gluing, doping, or painting is required. The wing covering and rudders are pre-built and all tubing is anodized.

"The airframe kit is ready to ship two weeks after we receive your order. The powerplant and landing gear kit go out in three weeks. The custom sail requires five weeks—between Pterodactyl and Manta we've got the sailmaker pretty busy! When you design the colors of your wing covering, use blue and green sparingly if at all. Yellow and orange are the easiest colors to see."

Pterodactyl Fledgling is priced at $2750, including the complete aircraft and Pterodactyl X powerplant. The landing gear uses shock cord suspension and 16-inch bicycle wheel mains to permit operation from unimproved fields. The powerplant uses a 242cc snowmobile engine with a direct-driven 36-inch propeller, quieter than the earlier reduction-drive 136cc Chrysler engine. See Table 6-2.

The Mitchell Wing

The top of the line among flying wing, powered hang gliders is the Model B-10 Mitchell Wing and its successor, the new Model U-2 Super Mitchell Wing (Figs. 6-13, 6-14, 6-15,15 6-16, 6-17) designed by aerodynamicist Don Mitchell. Flying wings go back to the 1930s when Mitchell first become interested in them. They have included Jack Northrop's YB-49 bomber design and such postwar concepts as the SV-45 Fauvel that first flew in 1950. Mitchell got busy on the concept again in 1974 when he got a call from a hang glider enthusiast, Dr. Howard Long, who wanted something with top performance. The result was the Mitchell Wing hang glider, which George Worthington flew to a world distance record and Brad White flew to win the 1977 U.S. Hang Glider Championship. In 1978, Steve Patmont flew a Mitchell Wing at Perris Valley Airport in Southern California and later flew it to win the World Ultra-Light Power Meet at Anoka, Michigan.

Table 6-2. Pterodactyl Fledgling Specifications.

Span	33′
Area	162 sq/ft
Chord at Root	5.5′
Tip Chord	4.5′
Sweepback	18 degrees
Dihedral	6 degrees
Aspect Ratio	6.8
Empty Weight	125 lb.
Max Gross Weight	350 lb.
Fuel Capacity	2½ gallons
PERFORMANCE	
Lift to Drag Ratio	9:1
Sink Rate (180# pilot)	250 fpm
Climb Rate	300 fpm
Stall Speed	–20 mph
Cruise Speed	25-45 mph
Top Speed	50 mph
Assembly Time (No Tools)	20 minutes

Fig. 6-14. The B-10 Mitchell Wing has set several world records.

Design of the B-10 is straightforward. It has a 12-degree sweepback, the center section is flat and the outboard panels have a 6-degree dihedral. The wing tapers in chord from five feet at the root to two feet at the tips. Span is 34 feet. Airfoil section is the NACA 23015. Standard controls are used with the ailerons doing double duty as both ailerons and elevators, or stabilators. Drag rudders are installed at the wingtips and act independently or can be used together as drag brakes.

The wing has no reflex. The stabilators are set at a positive 4 degrees with travel only up to 35 degrees, flying in their own airflow. The result is a perfectly pitch-stable wing. It has been static tested to 1250 pounds and the design load limit is over 1900 pounds, positive and negative.

The webbed spars have tapered spar caps, with D-section construction utilizing a 1-inch foam former every 4½ inches, covered with one mil plywood. Ribs are of standard truss construction covered with Ceconite. The wing itself weighs 70 pounds and the

Fig. 6-15. The U-2 Super Mitchell Wing has an enclosed cockpit and folding wings.

Fig. 6-16. Dick Clawson of Visalia, California, took this photo of himself in steep climb in a Mitchell Wing.

foot-launch hang cage weighs 11 pounds. The cage, with a McCulloch 101 engine using direct-drive and one gallon of fuel, weigh in at 47 pounds. In the "airplane" configuration, the tricycle gear, engine and gear-reduction unit swinging a 42-inch propeller and 2 gallons of fuel, weigh 150 pounds.

The Mitchell Wing can be assembled for flight in 10 minutes in any of the three configurations. Any of the cage arrangements are attachable with four pins. Assembly is simple. Remove the wing from the top of the car by removing the four pins and two cable tiedowns. Set the wing on the flying cage and stick the pins back in. Then unfold the wings and install the tip rudders and pins at the wing joints. After a thorough preflight safety check, you're all set to fly.

When rigging the wing, says James Meade, general manager of the Mitchell Aircraft Corporation, the stabilators are set to a posi-

Fig. 6-17. Dick Clawson loves to buzz along rivers in his Mitchell Wing.

tive setting of 4 to 36 degrees. With no negative setting, the wing washout prevents any tendency to tuck under. This rigging also results in the wing's center section always stalling first. This leaves the outer panels still flying and provides solid roll control at all flying speeds.

Seated snugly in the trike gear cage, the pilot has complete control in all three axes with no need to use body swing for either pitch or yaw control. The throttle is operated with the left hand, the right hand manages the stick control and the feet work the rudders either independently or together.

The engine instrument panel, mounted to the pilot's left, includes cylinder head temperature, tachometer, ignition switch and throttle.

The Mitchell Wing comes in kit form with the spars, fittings and stabilator D-sections complete. An assembly manual is included with step-by-step directions.

All parts of the Mitchell Wing kit are precut and the drawings are full-size. Wood-to-wood bonding is done with aircraft epoxy. Dacron covering is supplied but you buy the dope. All you need for tools are a one-fourth-inch hand drill, pop-rivet gun and such hand tools as a screwdriver, a hacksaw, a crescent-wrench pliers and C-clamps.

Mitchell Wing kits sell for $2700 complete. Several hundred kits already have been purchased. Close to 100 Mitchell Wings are now flying.

The popularity of the Model B-10 Mitchell Wing led the company to get Don Mitchell to develop a high-performance version. The Model U-2 Super Mitchell Wing was introduced at the 1979 EAA Fly-In at Oshkosh. Using a 125cc engine, the U-2 attains a cruising speed of better than 60 mph and stalls under 25 mph. It gets off in 150 feet and lands in 75 feet. Weighing under 140 pounds empty, the U-2 is stressed to more than 10 G's. The cockpit area is enclosed.

Some modifications are being made to the initial U-2 design. Included are different tips, with rudder control by spoilers located aft in the outboard wing panels. A fully retractable tricycle gear is used, the nosewheel is steerable, with brake. A swing-arm control stick has been added for comfort and ease.

For details write to the Mitchell Aircraft Corporation, 1900 S. Newcomb, Porterville, CA 93257.

Chapter 7
Flying Wings

Dale Kramer, 21, was a university dropout with a major goal in life. His explanation:

"I had two and one half years of education in aerospace engineering at the University of Toronto, but I left when I saw it wasn't leading where I wanted to go. My goal is to be self-employed in a business that will let me stay close to flying. Another year and a half has gone by since then and I hope I'm on my way!" Some education obviously did rub off on Dale. With his youthful enthusiasm and imagination, he and a friend, Peter Corney, have become real trailblazers in the fascinating world of ultralights.

The Lazair

There's no formal classroom in the high sky. No blackboard jungle substitute for trial and error development of a fresh idea. Instead of sitting through dry lectures about theories of LD and all that stuff, Dale and Peter hammered out a brand new design and had it flying in just two months. *Lazair* (Fig. 7-1) made its first public appearance at the Lakeland Sun 'N Fun EAA Fly-In in Florida in 1979.

Lazair is a neat little 136-pound, twin-engined, fixed-wing, inverted-V tailed ultralight that shows some brilliant insights in design and engineering. It is a growth machine that is not rigidly fixed in concept. The two Canadian youths aren't the kind to sit back and enjoy their initial success. The challenge of the future, as they say, lies ahead.

Lazair actually is Kramer's second design. His first was a fine little flying wing. Even before that Dale had built a Super Floater from a Klaus Hill design. Dale's Super Floater was built after he

Fig. 7-1. Canadian designer Dale Kramer's lively Lazair has an inverted V-tail.

bought a set of plans at Oshkosh '77. Believe it or not, two and a half weeks later it was flying.

Kramer's Super Floater was all that the name implied. He and Corney auto-towed it off his brother's 2000-foot grass airstrip. On its first flights it had soared up to 600 feet. It was difficult for Dale to keep his mind on university lectures. Three months later he knew he had to get busy on a design that had loomed big in his mind—a flying wing ultralight.

Dale and Peter worked long hours to get their flying wing ready to show off at Oshkosh '78, but time just ran out. It was just as well—at Oshkosh they were exposed to all the latest good stuff in the booming ultralight movement and their heads were crammed with new ideas.

Back home in Port Colborne, Ontario, they completed their flying wing. It was a thing of real beauty. Its fully cantilevered wing had a thick airfoil and was slightly swept forward to keep the center of gravity at quarter chord position. The pilot is positioned up front, where he could foot-launch with a running jump. A PVC covering was taped onto the wing, which featured tip rudders, with pitch control supplied by body shift.

"It has flown well as a glider," says Dale, "and I still believe in flying wings."

However, the day of powered hang gliders had dawned at Oshkosh, when John Moody unveiled his powered Easy Riser with a Mac 101 engine.

At Oshkosh '78, Dale bumped into Ed Sweeney, the Reno, NV designer of the Gemini twin-engine power pack for PHG's (Fig. 7-2). Sweeney had come up with an unusual innovation in power packs by linking two 5½-horsepower modified Swedish A. B. Partner chainsaw engines that direct-drive two-bladed propellers at

5000 rpm. The rig weighs 26 pounds and bolts directly onto the cage of any rigid-wing or flex-wing hang glider.

"After Oshkosh '78," says Dale, "Sweeney dropped in on us on his way to the East Coast. He flew my wing in its test rig, but when he saw the Super Floater (Fig. 7-3) he said, 'My engines will fit on that!' "

A Powered Super Floater

In no time at all, they had the Gemini engines installed on the Super Floater. Since then the flying wing has been in mothballs. The Super Floater with power had a real potential, says Dale, and it was flown a total of 15 hours before Sweeney left. Dale recalls: "It was great fun, flying just above the water along the shore of Lake Erie, practically walking on water and still knowing that if one engine quit, we could still make it to shore!"

As it turned out, Sweeney went back to Nevada with the Super Floater and left them with the two engines to install in a new design they hoped would get rid of all the faults they found in the Super Floater. In a bit over two months, Peter and Dale had a new machine flying—the Lazair. There were still many changes to be made to the engine mounts and the control linkages, varying from (would you believe) hydraulic cylinders to spring-loaded controls, before they decided to go back to pushrods and cables.

"I also went through five different engine mount designs," Dale says, "and with the Lakeland EAA show only a month off, we had yet to build a trailer."

Build the trailer they did. By virtue of working almost around the clock, they finally shoved off and arrived at Lakeland three days before the midwinter show opened. They'd flown Lazair only four

Fig. 7-2. This Lazair ultralight uses a Gemini twin engine powerpack designed by Ed Sweeney of Reno, Nevada.

Fig. 7-3. This Super Floater is an earlier Klaus Hill design.

hours prior to leaving Canada and then put on another two hours in Florida by the time the festivities began. When the show was over they'd run the total time up to 11 hours.

They didn't feel it was time to freeze the design yet, as they felt a single-engine version might be better than a twin. Once they settled on the final design, they would decide on kits and prices. Initial plans were to supply the leading edge D-cell complete, the ribs capped, and all tubing cut and bent. The only tools required for assembly would be general shop tools.

The Lakeland demo flights taught Dale and Peter a few things. Three noseovers occurred on ground runs and resulted in some broken tubes. "They were caused by two things," says Dale. "Lack of a nose skid for one thing. If the aircraft hit a ground obstruction it stopped abruptly, causing the pilot to swing forward and go through the front tube. The solution was to add a nose skid and pilot restraints. Another addition will be larger wheels. We've been thinking of using bicycle wheels for a smoother ride over rough ground."

Most of the Lazair development time was consumed by perfecting a control system, starting off with the hydraulic lashup they quickly discarded. From their experience in flying the Super Floater, they decided against using the rudder/elevator only system. This system had a tendency for the craft to slip in entering turns and Dale and Peter, both licensed power pilots, instinctively tried to correct by coordination.

As a result of their dissatisfaction with the Super Floater control system, they went to ailerons on Lazair—retaining all control functions in the stick. This required some fancy mixing. Dale's final version of Lazair ties rudder and ailerons together for coordinated turns, while retaining separate elevator control for pitch. You can't cross-control rudder and aileron to execute slips, which also makes crosswind takeoffs and landings memorable experiences.

The control mixer unit is installed in the wing root above the pilot's head. It is actuated by an overhead stick linked to pushrods, mounted in PVC guides, running back to the inverted-V ruddervators and through bellcranks and pushrods to the ailerons.

The Lazair pilot sits in a sling seat. This permits the operator's legs to hang down for foot-launch. The seat is attached to a pilot cage of 6061T6 aluminum tubing that is held rigid with nylon plugs where the bolts pass through. The Gemini engines are mounted to each side of the pilot with the blades shielded for protection. Sweeney modified the standard A. B. Partner Swedish chainsaw engines by turning down the crankcases, reversing drive and flywheels, and attaching the propeller to the back end where the starter formerly hung. At 5500 rpm, the engines put out their maximum torque. The propeller tip speed is subsonic and delivers 38 pounds of thrust from each engine. This is sufficient to haul a 160-pound pilot skyward easily, if not thrillingly.

Rate of climb for Lazair is from 200 to 300 fpm and it stalls at 17 mph. Cruising speed is 35 mph and it gets off the runway (paved) in 200 feet. Fuel consumption is one U.S. gallon per hour or 35 mpg! Wingspan is 36 feet 4 inches, wing area is 140 square feet and when empty it weighs 132 pounds.

Lazair's D-cells are 7 inches deep by 10 inches wide with 4-inch rib spacing at the root and 6 inches at the tips. Skin is .016 of an inch 2024-T3 aluminum. The wing spar is a C-channel, .025 of an inch with U-channel caps. Ribs are held in place with vertical U-channels not glued to the skin. Fittings are 7075-T-6.

The rear ribs are one-inch polystyrene foam with aluminum caps, the fuselage is all 6061-T6 tubing except for the boom.

Fig. 7-4. The Weedhopper uses a Chotia 460 single-cylinder powerplant with tuned exhaust.

Covering is 2-mil Mylar taped on. Two people can cover both wings in under six hours.

Weedhopper

John F. Chotia began building model planes a quarter century ago at the age of nine. He was the terror of the school playground with a special combat flying wing model powered with his own specially modified engine. He made many of the components himself. By 1964, he'd designed a man-powered aircraft which he claims was a dead ringer for Dr. Paul MacCready's prize winning Gossamer Albatross which flew the English Channel. In 1965, he was interested in hang gliders, half a decade before they caught on in America.

Like other glider-riders, Chotia held the dream of building something a little better in the ultralight aircraft field. It wasn't so long before he had turned out no less than 24 full-size ultralights. At that time he was working a four-year apprenticeship as an experimental machinist at NASA/Ames Research Center. By late 1977, he'd moved to Utah, a state noted for its soaring and ultralight flying machine activities.

Ultimate Fun

Quite naturally, he set out to design and build what he called the ultimate fun machine. And he came pretty close with Ultralight Number 22. He admits it flew only marginally, powered with a Chrysler 820 outboard motor and a 4:1 cog-belt reduction unit that swung a 45/12 propeller. He switched engines and installed a 242 Chaparral snowmobile powerplant.

By installing a tuned exhaust pipe and finely tuning the engine, he managed a few good flights. Number 22, had a double-surface wing span of 22 feet 8 inches and held 190 square feet of sustaining surface. There were no ailerons, just a rudder and elevator, and he rode in a reclining seat mounted on a tricycle gear. On to Number 23, where Chotia went to a single-surfaced wing. It was strictly a flying wing configuration which, he discovered, had excess control drag in tight turns. This was attributed to an excessive pendulum effect. So in Number 24, the present Weedhopper, Chotia added a tail group and designed a new fuselage.

Then Chotia began a search for a good engine and propeller. But none seemed quite right to him. He tried the Yamaha 292 and the Chaparral 242 with comparable results. And he logged some 80 hours flying time with them. The engine was mounted up front in tractor fashion. This is unlike most other powered hang glider installations. It permits the pilot to sit back on the center of gravity with the engine weight keeping it in trim.

The Chotia 460

Chotia saw the handwriting on the wall. People were buying up all available surplus snowmobile engines and he already disliked the high rpm inefficiencies of the Yamaha and Chaparral powerplants. So, being a trained machinist, he designed his own engine—the Chotia 460 (Figs. 7-4, 7-5, 7-6). It is the world's first powerplant specifically designed for use in ultralights. It delivers 18.5 horsepower at 3500 rpm, permitting use of a direct drive propeller with larger blades turning at half the speed of the Yamaha and therefore with higher efficiency. Efficiency is from 50 percent to 55 percent or 9.25 horsepower actual thrust.

Weedhopper's basic structure is of seamless drawn 6061 aluminum tubing that is reinforced at all attachment points with larger tubing or with wooden dowels. Pre-machined brackets or gussets are used to join the tubes. All bolts and hardware are of aircraft quality.

Covering is 3.8-ounce stabilized Dacron sailcloth, pre-sewn to slip into place, with no additional sewing, gluing or doping necessary. You have a choice of mixed colors including black, white, red, orange, gold, yellow, two shades of green or blue, and purple for a touch of regalness.

Completely assembled, Weedhopper weighs 160 pounds. Its 28-foot wing with 168 square feet of area provides a wing loading of around .95 pounds per square feet. Where other flying wing ultralights, such as the Mitchell Wing and Striplin FLAC, use elevons

Fig. 7-5. John Chotia's single-cylinder engine with two types of propeller blades.

for pitch and roll control in combination with tip rudders, Weedhopper has dispensed with any control surfaces on the wing. It relys on a rudder-induced yaw and wing dihedral to achieve a positive roll force and coordinated turns. A side stick is linked to the elevator on the tail for pitch control.

Ground handling is easy due to the wide tricycle gear's track and the low center of gravity. Optional cast aluminum rudder pedals or the standard rudder bar welded to the nosewheel yoke provide positive steering through the nosewheel. This is useful in making tight turns on the ground, as in parking. Due to the small amount of weight on the nosewheel, the rudder is the only control required during the takeoff roll. Incidentally, using the nose fork steering, the pilot resorts to "bobsledding." But with the rudder pedals installed, normal aircraft-type rudder use is achieved. This permits faster transition for experienced pilots.

The pilot reclines comfortably in flight with good forward visibility. Goggles are recommended due to prop wash on takeoff. In level flight attitude, the wash passes above the pilot's head. According to Chotia, a stall warning comes at roughly three mph prior to actual wing stall—with slight elevator buffet. Power-on stalls in straight flight are mushy, with minimum altitude loss of 10-15 feet. A power-off stall loses about twice that amount of sky.

In a banked turn in excess of 15 degrees, the inside wing drops and a sidelip to the inside ensues. This increases the angle of attack on the inner wing. Therefore, the wings level on their own. Spin testing had not been completed at this writing, but, says Chotia, Weedhopper appears to be spin-resistant.

The inherent stability of Weedhopper comes from its wing dihedral and low center of gravity. Overly sensitive control forces and power changes do not noticeably affect trim, says the designer. Weedhopper can carry a maximum load of 220 pounds at a redline speed of 50 mph or cruise at an easy 30 mph. With a 160-pound pilot aboard, it stalls at about 22 mph. A standard one-gallon plastic fuel tank will get you over 30 miles with the Chotia engine.

With a 190-pound pilot aboard at Chotia's home base in Utah, at 4500 feet elevation, Weedhopper demonstrated a takeoff capability at 190 feet on a 90 degrees Fahrenheit day. Rate of climb was in excess of 300 fpm. Its short-field capability and tight maneuverability make it an ideal off-airport craft, Chotia says.

He describes flying the Weedhopper this way: "Imagine yourself floating through early morning mists, climbing out over the trees, soaring along a ridge, engine off and with only an eagle

Fig. 7-6. John Chotia's Weedhopper ultralight has a ruggedly simple design.

for company. Picture yourself skimming along over open fields on a bright summer day, out in the open, low enough to enjoy the scenery—just like flying should be!"

If Chotia sounds a bit poetic, he is also a very practical person. Consider the development of the little Chotia 460 engine. He says: "Its performance is outstanding, with over 650-fpm rate of climb at sea level. Add to this a lower noise level, lower fuel consumption, lower vibration levels, comparable weight to other ultralight powerplants and it is easy to see why we decided to build our own engine."

The Chotia 460, he says, "is much lighter for its displacement than normal, high-horsepower-per-cubic-inch two cycle engines. The low rpm means longer life and we expect a TBO to run between 800 to 1000 hours. Timing is adjustable from the cockpit, so you can retard it for easy starts and smooth idling and peak it for maximum performance simply by watching your tachometer."

The crankshaft is extended and a third ball bearing is added in the extension case to carry the propeller loads. The direction of rotation is reversible and the engine is easily adaptable to other ultralights.

In designing the Chotia 460 engine, care was taken to eliminate excessive vibration common to single-cylinder powerplants. Piston weight was considered the critical component determining mass vibration. Therefore, Chotia made the piston very lightweight, at 12.5 ounces, with an 88mm bore. Therefore, a lightweight connecting rod was possible. It was machined from bar stock 7075-T6 aluminum and fitted with needle bearings and hard steel braces at each end. A low 5:1 compression ratio also was used to minimize vibration, providing a soft "push" on the power stroke.

Chotia also decided to design the low-rpm two-cycle engine with small ports for good ring support and mild timing for good low-speed torque. This, coupled with the rod needle bearings and triple ball bearing crank, is expected to yield a long life. The crankshaft is of 30mm diameter, heat-treated 4340 steel.

"One thing I added while making the patterns," says Chotia, "was the use of four bolts to hold the exhaust flange rather than the normal two, to avoid the possibility of them shaking loose."

On a recent visit to Southern California, Chotia put on some demonstration flights at Santa Susana Airport for a TV crew brought in by commentator Arthur Godfrey. Godfrey is a 17,600-hour ex-Navy pilot and the film was for a national spot. As it happened, a cold front was approaching that day and Chotia demonstrated Weedhopper's ability to take off and land in a 10-knot direct crosswind. It appeared to fly well in moderate turbulence at pattern altitude. Banking and turning was done gracefully over the field as he skimmed a nearby grassy hillside.

At rest, Weedhopper presents an unusual appearance. Its wing sail droops like wet laundry on a clothesline. The moment it lifts off and total circulation fills out the wing, it becomes transformed into a thing of beauty that is eager to play with the birds and explore the countryside low and slow—the essence of natural flight.

Noted test pilot Bob Hoover flew Weedhopper on September 8, 1979, at Page Field, Oklahoma City. He did no aerobatics, but he did land with a big grin. The ship had been built by Reklai Salazar and was the 79th Weedhopper to fly of more than 300 kits sold at that time.

Weedhopper kits sell for $2495, including all tubing, hardware machined components, wheels, engine, propeller and controls. The engine is available separately. Optional equipment includes a nosewheel fender, upholstered seat, 3.5-gallon fuel tank, storage bags and a double-surface wing for higher performance.

Weedhopper is considered a true airplane by the FAA and not a powered hang glider. It uses a conventional tricycle gear for takeoff and landing maneuvers. A Weedhopper pilot must obtain a second-class medical certificate and hold a student pilot license to fly it legally.

Kits are available from Weedhopper of Utah Inc., Box 2253, 1965 S. 1100 W, Ogden, UT 84404. Information on kits are available for $5.

Chapter 8
Powered Sailplanes

When the Westerly slope winds blow up and cross the awesome Wasach Range east of Ogden, UT, things start to happen at Morgan Municipal Airport. The airport is nestled in a lovely valley at 5020 feet elevation, far below the 9000-foot peaks surrounding it. Winter or summer, hangar doors slide open and strange looking ultralight homebuilts emerge looking somewhat like happy dragon-flies.

Mountain Green Ultralights

Morgan is home base for the late Klaus Hill's Mountain Green Sailwing operation, a sort of three-man outfit that specialized in designing and building flying machines that weigh barely more than do their pilots (Fig. 8-1).

Environmentally compatible with their surroundings, these lovely craft mark a new departure from traditional homebuilt design and construction. They are not meant to go fast. Rather, they perform like graceful butterflies in a summer zephyr, flitting from peak to peak on the free energy of the wind and sun.

The Mountain Green craft are similar to, but not quite like sailplanes or hang gliders, two related movements that have given special impetus and meaning to the ultralights.

Visitors to the EAA Fly-In at Oshkosh in the summer of 1977 first saw Klaus Hill's foot-launched Super Floater, and searched in vain for the location of the ailerons. There were none. Super Floater is strictly a two-control machine. It utilizes rudder and elevator only and it is controlled by a single side stick.

Fig. 8-1. Mountain Green Sailwing's Humbug at the Oshkosh Fly-In in 1979.

A high performance aircraft, it proved easy to build in under 500 man-hours, including design time. The materials cost only $545. Performance worked out at roughly 14:1 LD at 26 mph, minimum sink 3.0 ft/sec at 23 mph, and stalls under 20 mph.

There was no big rush to buy plans. A foot-launched ultralight then seemed just a bit too new for most EAA members to grasp. But in Morgan country, the winds were right and a whole new movement was soon to be born—one apparently destined to go places.

In the Mountain Green ultralights, Hill gave you the option of taking off by running into the wind and landing on a skid after a comfortable flight sitting down chasing thermals. A step beyond hang gliding, for sure, but still not the ultimate design.

The Honeybee

So now let's look at a couple of newer Morgan ultralights—the Honeybee, designed by Roland H. Sinfield, and the Hummer, another Klaus Hill production number. First to fly was the Honeybee (Fig. 8-2) described by its designer as half airplane and half hang glider, with signs of being part snowmobile. Wing construction is cable-braced aluminum tubing with a single-surfaces airfoil. Tail surfaces are of aluminum tubing with fabric cover. The fuselage is built of four wooden stringers covered with thin plywood. The result was a bunch of nicknames like Preying Mantis, Flying Railroad Tie, Pogo Stick and Flying Pencil.

"Main design goals were simplicity of construction, low speed, low cost and something that would give the pilot the same thrill we all got on our first airplane ride," says Sinfield. All four goals, he says, were achieved nicely—especially the last. "Also," he adds, "after several modifications, the handling characteristics are better than we had hoped for."

Since the craft is mostly tubing, cables and bolts, it can easily be broken down into small components for transportation or storage. For transporting only small distances, you simply remove the wings. Only two parts require outside fabrication—the welded landing gear and the sail (cloth portion of the wing), built by a professional hang glider manufacturer using a heavy duty sewing machine.

The control system, like Super Floater's, is rudder and elevator only. Sinfield says it is plenty adequate for a slow ultralight. "With the rudder mounted low in relation to the longitudinal center of mass and the fact that the wings have considerable dihedral," he says, "turns are surprisingly coordinated, not skidding as you might expect."

Initially, they connected the rudder to foot pedals. Later the hookup was changed to link the rudder cables to the control stick, which is moved sideways in turns. The result, says Sinfield, is a more natural feeling for experienced pilots.

Crosswind landings might be a problem. In that case, the pilot simply moves off to an open space and lands into the wind rather than using the wrong runway. Morgan Airport has only a single gravel strip, 3800 feet long—Runway 3-21.

Powerplant is a single cylinder 395cc JLO engine, which Sinfield feels develops power at a lower rpm than most two-cycle engines, providing greater propeller efficiency. It also has a high power-to-weight ratio and the cost is low. He says you can scrounge many out-of-production snowmobile engines at around $100 to $250. Although Honeybee's engine has performed well so far, the JLO is a real Shaky Jake due to its large single-cylinder design.

Sinfield expresses some concern over the rear mounting of the pusher-engine configuration. In the event of catastrophic engine failure, the small fuselage could be sliced off. For this reason, he recommends using a husky tapered crankshaft and a strong propeller and prop hub.

There are other pros and cons in the pusher arrangement. It severely limits the CG range fore and aft. Yet, as Orville and Wilbur learned, the propeller efficiency is improved because the air it thrusts rearward doesn't have to flow over the whole craft. The tail, however, must be built ruggedly to withstand the higher blast forces of the airstream.

Honeybee is not exactly IFR equipped. It carries a hiker's altimeter and a Hall windmeter, mounted to the pilot's right knee with a Velcro band to avoid engine vibration.

Fig. 8-2. The Honeybee is another Klaus Hill design from Mountain Green Sailwings in Morgan, Utah.

In its first winter of flight tests, between snow storms, Sinfield got Honeybee up to 7200 feet and logged close to 10 hours flight time. Not quite enough time, he feels, to offer plans for sale. But sufficient to offer an information booklet showing construction details, component photos, address of engine, propeller, and other parts suppliers, plus updating on Honeybee—which by now could well be on its way as one of the newest of the breed. Send $5 to Roland Sinfield, P O Box 513, Morgan, UT 84050.

The Hummer

The second new ultralight to appear at Morgan was the Hummer (Fig. 8-3), Klaus Hill's newest design. Klaus started building primary gliders in Germany 25 years ago. Among his earlier craft were his Hobby sailplane, Fledgling hang glider, and the Super Floater.

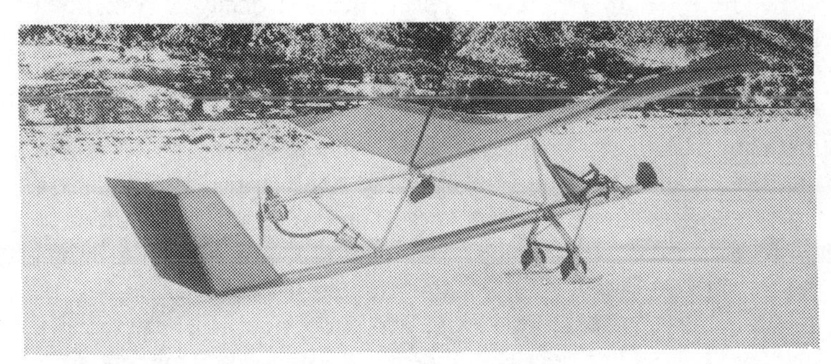

Fig. 8-3. The V-tailed Hummer is a graceful ultralight designed by the late Klaus Hill.

Hill got busy designing Hummer following successful flight testing of Honeybee, which he worked on with Sinfield and Larry Hall. In just three weeks he had Hummer humming all over the mountain valley at Morgan, where it generated so much interest that a number of local pilots ordered kits. See Table 8-1.

Similar in construction to Honeybee, Hummer has a double-surface wing, a V-tail, aluminum tubing fuselage, smaller, smoother engine, a recoil starter for air restarts and an instrument panel with a tachometer, ROC, airspeed and altimeter.

Powered with a 22 horsepower Chaparral snowmobile engine, Hummer turns in considerably better performance than the lower-powered Honeybee. It has a faster climb, better LD (10:1), lower sink rate, better penetration, less drag at high speeds and less engine vibration—about like that of a J-3 Cub.

Klaus followed up his initial flight testing of Hummer with a production run of 10 kits to be sold at $1800 each, complete with engine and propeller. Hard-to-make parts are prefabricated, leaving only the simple operations like drilling, pop-riveting and rigging to the homebuilder. He felt that future kits will hold the price at that level.

Says Sinfield: "We are continually learning new things, such as the large changes in aircraft performance that result from small changes in propeller design. Also, the greatest difference between flying small airplanes and flying ultralights is the latter's unusually slow airspeed. The first few flights may be shocking to the average lightplane pilot, yet it is just this very slow flight speed that makes these craft so enjoyable. Flying ultralights in turbulence is a challenge, one we do not recommend attempting."

It was not a matter of turbulence, but of an apparent pitch instability that took the life of Klaus Hill on a test flight of a new ultralight, the *Voyager*, on October 2, 1979. Originally, Voyager had been built as an oversized hang glider with 40-foot wings for a 250-pound pilot who was not satisfied with its performance. Hill took it back and began modifying it into a powered hang glider by installing a Chaparral engine in tractor fashion. He put wheels on the control bar and a tail wheel behind. It was similar in appearance to a Pterodactyl and in taxi tests it worked fine. A harness was added to protect the pilot from lunging forward into the propeller in case of a crash landing. Finally, Hill decided to take it up.

The first flight, a low sortie, was successful. However, Hill's partner, Larry Hall, recalls it appeared to be a bit "squirrely." On the second flight, Hall remembers, Hill took off and climbed to

Table 8-1. Hummer Specifications.

Wingspan	34′
Length	17′
Chord	51″
Wing Area	134 sq ft
Engine	22 hp Chaparral Snowmobile
Propeller	Klaus Hill Special
Top Speed	50 mph
Cruise Speed	35 mph
Stall Speed	24 mph
Empty Weight	170 pounds
Gross Weight	340 pounds
Wing Loading	2.6 lb/sq. ft
Power Loading	2.6 lb/sq. ft
Load Factor	+6 G's calculated
Climb	250 fpm (at 5000′)
Ceiling	8200′
Fuel Consumption	2 gph

roughly 200 feet altitude. It pitched down, recovered, then climbed back and repeated the pitch oscillation in roller coaster fashion. Hill apparently cut power at one point, then returned to full power in an effort to stabilize the craft. On the third dive, the Voyager did not recover—it struck the ground, killing the pilot.

Meanwhile, Hall elected to carry on the Mountain Green Sailwing operation by producing Honeybee plans and kits. He can be contacted at Box 771, Morgan, UT 84050. The Hummer, says Hall, was taken over by Dennis Franklin of RR 2, Glen Rock, PA 17327.

The Backstrom Flying Plank

For the past quarter-century, a talented aeronautical engineer named Al Backstrom has followed a dream—to design, build and fly the ultimate inexpensive, easy-to-fly, self-launching sailplane. Now that powered hang gliders have come along, a fresh look at the Backstrom WPB-1 Plank (Fig. 8-4) is in order.

What Backstrom had in mind was not one of those things you tuck under your armpits and rise to the occasion by running and jumping into the air. The Plank is much much more.

You sit comfortably in the Plank's reclining seat inside a smoothly contoured cockpit. In Backstrom's dream for a flying Plank, a tricycle gear with permanently retracted mains, like in other sailplanes, makes ground handling super easy. A pusher propeller behind the tailless flying wing is driven not by rubber bands, but by any one of many available two-cycle engines in the 10-horsepower range such as the Chrysler West Bend or the McCulloch 101.

Backstrom, a veteran sailplane driver, has described his concept for a Self-Launching Plank Sailplane (SLPS-1) as his personal solution to lower cost soaring with a light weight, semi-homebuilt design that eliminates the expense of hiring a tow plane.

Self-launch also is made practicable by foot or hand launch from ridge sites, he says, but he prefers the powered self-launch concept where you drive off level ground into the sky in maybe 500 feet (sea level) and climb like a homesick angel.

Described as phase II of the Auxiliary Powered Plank Sailplane design, this version would have a span of 34 feet 8 inches and a constant chord of 4 feet for a wing area of around 138 square feet. It would have a length of 9 feet 6 inches, empty weight of 130 pounds and a design gross weight of 370 pounds using a Mac 101 engine. It would knock down into four pieces for trailing (three-piece wing plus the pod) and would be built from wood and PVC foam/glass sandwich materials.

Preliminary performance estimates by Backstrom called for a rate of climb of 400-500 fpm (standard day, sea level) and with the engine shut down an LD maximum of about 23:1 or over 25:1 as a pure sailplane minus engine and propeller. What makes the Plank different from other motor-gliders is its utter simplicity of design and ultra-light weight. Backstrom built his first Plank in the 1950s, with the help of two friends, Jack Powell and Phil Easley. In 1969, he undertook design studies for a powered version. Construction got under way in 1972 with the main work done by Van White, an EAA official from Lubbock, TX.

Initially, Backstrom and White intended to install a Sachs Wankel or OMC snowmobile engine in the 20-horsepower range. But when none was available, they settled for a Kiekhaefer Aeromarine 440 fan-cooled, single-ignition, reciprocating powerplant.

"The recip feature sort of did us in," Backstrom said. That was because such engines require shock mounting instead of hard mounts like the Wankels use. The Kiekhaefer also was bigger and too heavy.

White worked on the powered Plank for about two years, doing the entire construction job on a 4 x 8 foot plywood work bench. Initial taxi tests and short flights were made at Lubbock, where they found the power output critical. By means of some fancy exhaust tuning, they got the rpm up to 5400. However, they found the Lake injector carburetor required considerable adjustment.

More recently, Backstrom has installed a conventional carburetor. With a geared belt drive system, he got 6500 rpm at the

Fig. 8-4. Al Backstrom's Flying Plank has a sleek look, an enclosed cockpit and wingtip controls.

engine and 2500 rpm at the prop—a two-bladed, wooden 54/52 club.

Backstrom designed his own airfoil. It featured a rather heavily reflexed trailing edge, a steep curve at the leading edge and a thickness of 15 percent that was excellent for low speed flight.

When he first started working on the Plank design, Backstrom took the position that something radically new had to be done to get away from the early standard homebuilt designs which he called "strictly adaptations of the simple light plane designs of the Twenties and Thirties, modified for a 65-horsepower engine."

From there, Backstrom said, EAA's movement has progressed from the early designs to sophisticated craft "that rival the complexity of current production aircraft of the same weight category."

In April, 1973, writing in *Sport Aviation*, Backstrom expressed the thought that it was about high time that "some of us in EAA should look into the possibilities of really flyable ultralight airplanes. These designs should be based on currently available materials and equipment and not a rehash of what had been done in the past."

The challenge has been met in various ways with the proliferating number of ultralight designs, particularly those using modern composite construction materials and methods—fiberglass, Dynel cloth and styrofoam blocks.

But Backstrom stayed with aluminum sheet, steel tubing and Dacron fabric for covering in the Planks. He strove for ways to make individual parts serve more than a single function in order to save weight. Besides weight saving, his goal was to minimize drag, hence the flying wing Plank design.

"It would be of little use to build a light airframe and then use all the power available trying to drag it up to flying speed," he said. "Put your helmet and goggles back in the closet and forget the wind whistling in the struts and wires. Drag items such as these can't be tolerated in an airplane that is to fly on low power."

In the spirit of making parts do double duty, Backstrom went to elevons for pitch and roll control, and drag rudders on the tip fins for directional control. The tip fins were meant to provide some end-plate effect to increase the effective aspect ratio as well as provide better directional control stability.

Backstrom went to the constant chord wing because it was simpler to build than a tapered wing. However, it did give a weight a and drag penalty. After initially flying the Plank with two tandem wheels, he decided on a trike gear arrangement. He also contemplated changing the thrust line by lowering it since the effect of power changes was quite noticeable.

In 1976, Backstrom made initial flight tests of the Plank with the trike gear. At Oshkosh '77, he flew the Plank repeatedly in the flyby pattern and attracted much interest among the thousands of builders who had read about the Plank but had never seen it fly.

Next, Backstrom went to a two-place design with a staggered side-by-side seating arrangement to cut down frontal area. He felt the craft could easily tote two people. Unfortunately, Backstrom did not plan on selling plans or kits because he was an FAA engineering service representative for the Southwest region, it could be construed as a conflict of interest.

Therefore, he put the Plank design up for sale and looked for some group or individual to take over the design development and kit production for his SLPS-1 Self-Launching Plank Sailplane.

Simplicity of the design is evident in the specifications. The shoulder-mounted constant-chord wings have cambered tips and tip-mounted fins and rudders, with identical ribs spaced at 6 inch intervals. Leading edge is of molded plywood and the rest of the wing fabric is covered and glued. Wing control surfaces are elevons.

The fuselage is built of tubular steel and fabric covered with Dacron. Flush engine air-ducts are forward of the wing's leading edge. Gear is the non-retractable tricycle type and the powerplant is the Kiekhoefer Aeromarine 440, driving the two-bladed wooden propeller as a pusher. Plank has a span of 21 feet 8 inches, a constant chord of 4 feet 6 inches and a wing area of 97.5 square feet. Empty weight is 390 pounds, including a radio and battery. The battery is used for ballast in the pod.

Performance figures are: 109 mph maximum velocity, with a takeoff speed of 50-55 mph and a climb of 400-500 fpm SL. Plank has a charisma all its own and eye-stopper appeal that makes it not only the predecessor to today's ultralight movement, but suggests the shape of things yet to come in the future of the homebuilt action.

Chapter 9
Very Light Planes

It's been called the flying moped, an airborne snowmobile and a sky-high trail bike. It's so tiny its initial test pilot, 235-pound Lowell Ferrand, had to go on a diet to fly it. But it's one of the earliest and more successful of the modern day fleet of ultralights that sprang into existence after OPEC began sky-rocketing crude oil prices.

The PDQ-2

Long before aerial sports began hanging McCulloch chainsaw engines on their hang gliders, Wayne Ison's *PDQ-2* (Fig. 9-1) was buzzing around Elkhart, IN like a mad hornet. After a few years of showoff at Oshkosh's annual EAA Fly-Ins, it has been accepted as a tried-and-true, plans-built, do-it-yourself escape machine that gets you airborne for under $1000.

PDQ-2 is a sort of strap-on flying machine with shoulder high monoplane wings, a tiny tricycle landing gear and a pylon-mounted JLO snowmobile engine that swings a two-bladed pusher propeller. The propeller blasts air backward over a swept T-tail hung at the end of a boom borrowed from a Bensen Gyrocopter. A 2-inch square beam of one-eighth inch thick 6061-T6 stiffened with guy wires is used.

The pilot sits up front biting bugs at a maximum of around 80 mph, fast enough for sure for this kind of open-air aviating. PDQ-2 climbs about 400 feet per minute on a cool day. Every pound counts

Fig. 9-1. Pilot Bill Jones flies his PDQ-2 over the Mojave Desert.

when you're flying with a Rockwell JLO LB-60002 engine of some 40 horsepower or with an alternate VW engine.

A few PDQ-2 builders use off-the-shelf 6 gallon outboard motor fuel tanks slung beneath the pilot's seat. They are also eyeing such other alternate powerplants as the BMW motorcycle engine and the Hirth 280-R. "Any engine that can put out a minimum of 36 horses at 70 pounds weight would work fine," Ison says.

While you should not consider flying the PDQ-2 in a snowstorm or through a fogbank, it rates high as a VFR fun aircraft that could be classed as a minnie recreational vehicle of the air. Whatever you want to call it, PDQ-2 is a real flying doll, one I had the pleasure of photographing from my Cessna 170 as flown by a West Coast PDQ-2 builder, Bill Jones. It is not disrespect that caused Jones to sell his PDQ-2, as well as a fine little Pietenpol Air Camper, and switch to an Ercoupe recently. Fun is fun, but taking the family flying puts the little beauty in a different category from a family flivver.

Wayne Ison, designer of the original PDQ-2 who sells plans and information kits ($5) from 28975 Alpine Lane, Elkhart, IN, formerly was active in the go-kart craze. When that movement became too sophisticated and expensive, he found a way to get back to basics by forming a beginner's category called "West Bend Class—Bushings Only". That was like going back to the old Soap Box Derby days.

Success of the West Bend movement stuck in Ison's mind when he turned to flying for recreation. For a while, he flew a restored Rearwin. Then he started to build a Bowers Fly Baby, but neither effort really turned him on to a way to create a "West Bend of the sky."

He scrounged around his garage, and being a sharp mechanical engineer, started putting things together in his head. He had the completed wings of his unfinished Fly Baby—he'd sold the fuselage to a friend—and a VW engine in a corner. Ison got out his tools and began getting his act together. The wings were joined to a bare-bones fuselage consisting of a couple of booms that went back to the tail. He sat up front with the VW engine virtually in his lap. It was all great fun, driving it up and down the runway at Elkhart, and mushing along in ground effect. The machine was given a name—the PDQ-1.

Ison felt he was on the right track. By mid-1973 a second ship, PDQ-2, was finished. In a way it was even simpler than PDQ-1. There was only a single tail boom sticking back from a vertical mast and keel, which passed for the airframe. Everything else hung on it. He paid a professional welder $20 to join the tubing to make it lighter than if he'd used nuts and bolts and gusset plates. Cleverly, he designed it so that two frame members joined at every stress point.

The landing gear was designed around a chunk of 2024-T4 aluminum 30 inches long by 2 inches by one-half inch, bolted to the front end of the keel and carrying a short rudder bar at the front end. For a nosewheel, he scrounged a 6-inch stock aircraft tailwheel. The mains were a carryover from his early days—a pair of 5-inch go-kart wheels with 3.40/3.00-5 double ply tires.

Jack Cox, editor of the EAA's *Sport Aviation*, took one look and shook his head. He quietly suggested to Wayne that he glue a layer of brake lining to the soles of his shoes for ground control.

Ison next put the Fly Baby wings back in storage and designed a new set made from Styrofoam bonded over plywood ribs at the root and the tip of each panel, with another seven foam ribs in between. He provided a set of four spruce spars—a one-fourth inch thick leading edge spar, a one-half inch main spar and rear spar and a one-fourth aileron spar.

The foam sheets, three-fourths of an inch thick by 4 inches by 8 inches, were cut with a saw roughly halfway through to allow bending over the curved ribs, which were shaped to the NACA 63 2A 615 airfoil. The foam was covered with Dynel, bonded at the edges and heat shrunk, with resin squeegeed into the cloth and lightly sanded. Microballoons were applied next and low spots filled with automotive putty before final sanding and painting.

The ailerons were full-span, of Dynel covered foam built around a one-fourth inch thick spruce spar with inboard plywood ribs and gussets. The tail feathers were similarly made.

A curved stick throttle attaches to PDQ-2's keel behind the vertical mast and ends where the pilot can grip a motorcycle twist-type hand throttle. It is spring-loaded to return the engine to idle if released. The complete power train, including the JLO engine and propeller, weighs 70 pounds. The direct-driven prop, a 44/17 design that delivers up to 180 pounds static thrust, was carved by Ison.

Lowell Farrand, the test pilot, flew PDQ-2 in ground effect until he got used to its feel. After a half hour's fun on May 1, 1973, he found himself flying—two feet off the deck. Initial runs proved the nosewheel steering was super-sensitive and the engine a bit under-powered. Ison redesigned the nosewheel steering.

"This redesign really saved me," says Farrand. "I was enjoying flying PDQ-2 so much on takeoff I almost ran out of runway to set her back down. So I'd slide around in a fast turn at the end of the strip. It was then I realized how good that gear really was!"

At this time, says Farrand, he flew with starter, flywheel, generator, exhaust stacks and a lot of other heavy items he didn't really need. "I was also overweight at 210 pounds and with my leather flying jacket, helmet, and so on, I weighed in at 225 pounds. We couldn't have that, so I went on a diet."

Later on, Ison lightened the engine weight and carved a new prop with less pitch. The power came up to where it belonged and Farrand was really eager to fly. "I didn't tell them I was going for high," he recalls. "But away I went. Then—shortly after leaving the strip, there was a *bang*! The linkage to one of the two carburetors came off, went through the prop and down through one fuel tank into the wing. The engine kept on running smoothly, but the rpms dropped off with the one carb out.

"I was over a mile-square cornfield, but I could see a newly cultivated bean field. If only I could reach it! There was a row of trees to get across, but I found a low place between two trees and lined up for a landing with the rows. I was soaked in gasoline, which was being sucked out of the top of the wing right onto my back. But I made it across the trees and landed okay."

The PDQ-2, he says, "is one airplane you can just pick up and carry back to the airport." With the carb linkage repaired, he was off and flying again.

Ison kept on working on the wing design. One winter day when the field was closed due to a blinding snowstorm, Farrand got the PDQ-2 out and flew it down the runway about 10 feet off the deck. He noticed something weird—the whirling snow flowed forward around his body and the center section and then back at a wide angle that put the wing in disturbed air. With this "flying wind tunnel" demonstration as a guide, Ison found a simple solution to the problem. He added inner end plates, or flow fences, at the roots and the flying characteristics improved dramatically.

On the ground, PDQ-2 handles easily in a crosswind because there's no fuselage surface for the wind to blow on. The gear works equally well on hardtop, grass, mud or snow, Farrand says. The open framework also makes visual inspection simple and should make a gyrocopter pilot feel right at home. The view from the pilot position is the same.

On takeoff, the nose comes up almost at the start of the roll and liftoff occurs at 35 to 40 mph. Then you fly level to build up speed to 55-60 mph and climb out. "The landing is the greatest," Farrand says. "You just throttle back a little and come down where you want to, then bring the nose up a little and its all so smooth you can't believe it. Again it's that wonderful gear—set her down crooked and whe straightens all by herself."

The high pusher prop, he warns, is a good brake. Don't throttle down too much in flight "or you'll kill your airspeed *now*. But it's fun to fly down into ground effect on your approach, then add power and fly it a few feet off the deck right up to where you want to park. It's all so slow you think you could step off and walk faster than you're flying!"

Bill Jones got into the act in 1974 when he bought himself a set of PDQ-2 plans and had it all together by September, 1975 when he made his first hop. He'd put on 78 hours of fun flying when I ran into him at Mojave Airport north of Los Angeles. A former Rogallo hang glider pilot, he'd simply decided he wanted to do more than just slide downhill.

Fig. 9-2. The Birdman ultralight was first designed by Emmett Talley.

Jones' checkout in the PDQ-2 was about like Farrand's first hop. There he was on a high-speed taxi run when all of a sudden he was airborne at the dizzy altitude of three feet. The Jones PDQ-2 has a wingspan of 20 feet, or 18 inches more than Ison's prototype. He has climbed to 5000 feet MSL over Mojave on a day when the density altitude was 8000 feet. But if he's abandoned the low-and-slow regime for a store-bought machine, don't worry—there's a plane for every pilot and vice versa. He's had his fun, and you can too, in a PDQ-2. Plans are available from PDQ Aircraft Products, 28975 Alpine Lane, Elkhart, IN 46514. See Table 9-1.

The Birdman TL-1A

The idea of flying like a bird has intrigued man since early times. Leonardo da Vinci was among the first to put down on paper the mechanics of bird flight, but only recently have ultralight aircraft designers and builders come close to realization of the old dream. One such effort is the Birdman TL-1A (Fig. 9-2). The story is related here by Jim Welling of Birdman Aircraft Inc., 480 Midway, Daytona Beach, FL 32014.

Imagine a quiet Saturday morning. The breeze is barely rustling the fallen leaves. The fields are ablaze with the rising sun and a lone dove silently and slowly glides along. As a witness, you slowly become aware of how awesome an experience this is. Your pulse quickens as you realize how soon you will cease to be a witness, but rather a participant.

Unhurriedly, you get out of your car and begin to assemble your Birdman Aircraft. About 30 minutes later the last bolts have been tightened and the safeties are in place. Leisurely you conclude your preflight and take your place in the cockpit. Satisfied

Table 9-1. PDQ-2 Specifications.

Span	18′ 6″
Chord	42″
Airfoil	NACA 63 2A 615
Wing Area	64.75 sq/ft
Wing Loading	6.5 lb/sq. ft
Span Loading	22.7 lb.
Empty Weight	218 lb.
Gross Weight	421 lb.
Top Speed	80 mph
Cruise	70 mph
Rate of Climb	400+ fpm
Stall Speed	46 mph
Engine	Rockwell JLC LB-600-2

with the response of your control surfaces, you set the throttle at idle and start the engine.

Advancing the throttle, the aircraft begins to roll. In a second, the tail is up and the sod beneath you begins to pass more quickly. At a trot, your aircraft lightens and as it gathers speed you can't keep it down. At 50 feet you begin a 180-degree turn. When you've turned completely you once again survey the fields.

This time the dove is not alone.

Dreamlike? Perhaps, at first, it might seem so. Yet this dream was transformed into a reality by Emmett M. Tally III. In 1970 Tally began the long and laborious task that was to end tragically for him on May 3, 1976.

Tally dreamed of creating a machine that would enable man to "fly like a bird." He felt that the world of aviation was lacking an aircraft capable of linking man and machine in a total expression of flight. By February, 1975, he had succeeded partially with completion of his first prototype, a T-tailed version of the TL-1. However, because of its size and sluggishness, it did not satisfy his dream.

Design and testing began almost immediately on a new prototype. This time the aircraft was to be ultralight and V-tailed. In December, 1975, the Birdman TL-1 made its maiden flight during what was meant to be a simple taxi test. The airplane yearned to fly and Emmett had fullfilled his boyhood dream. He had created a machine for the purist—for those who enjoy the wind-blown exhilaration of an open cockpit and for those who have always longed for the thrill and freedom of flying like a bird.

At the Sun'N Fun Fly-In at Lakeland, FL that winter, the Birdman TL-1 was an unqualified success. By April, 1976, Emmett had sold several hundred airplanes and was determined to show his airplane to the world. He journeyed to California where, on May 3, 1976, he departed Corona Municipal Airport in the TL-1 for the parking lot at the Anaheim Stadium. A big sport show was under way there.

Upon arrival, he circled the lot three times to the left, then reduced power on the last downwind leg. The aircraft initiated a left turn at about 250 feet. Loud snapping and popping noises were heard. The aircraft apparently inverted and pitched straight down.

National Transportation Safety Board accident investigators of the fatal crash concluded that the Birdman TL-1 had experienced structural failure of the left wing. Whichever came first, the failure of the leading edge or the tail-feather attach points, the Birdman was gone. And with it, Tally's dream.

Or was it? To revitalize an organization that had been inspired by a man with a dream would be a tremendous undertaking. A new prototype would have to be designed, tested, researched and developed. To be fair to those who had shared Emmett's dream, it would be necessary to provide well over 200 airplanes. However, the challenge was there and in June, 1976 the project began with vigor.

First, a replica of the TL-1 was constructed, then shaken, tested, broken and improved. Prior to any engineering changes, the airframe was dynamically tested at four times the amplitude modulation of our engine (14 hz-cps). The engineering numbers were run through computers and at the higher frequencies they occasionally approached yield points. But at no time did they near ultimate or breaking points.

Yet this would not be enough. To insure the safety of the airplane, there are now eight stringers running fore and aft—the entire length of the tail cone. At the attach points of the tail feathers to the tail cone there is now an external saddle patch which distributes the aerodynamic loads from the tail feathers over several bulkheads. Internally, bulkheads were added. Attached to the bulkhead, stringers, and skin there is now a three-fourth inch by 1 inch by 4 inch laminated buildup of sitka spruce. The forward tail-feather attach bolts now bolt through the buildup, skin and saddle patch.

The original aircraft had uncovered styrofoam leading edges. Therefore, it became necessary to find a way to protect this critical area. Two materials were discovered which, when applied to the foam, reduces all deterioration due to ultraviolet radiation. As a bonus, this covering also added increased strength to the leading edge and D-cell. Other substantial improvements include:

- A new tripod floating engine mount system which reduces airframe vibration by 98 percent.
- A new, spring-mounted landing gear.
- A new control box which combines pitch (elevators) and roll (sequentially activated spoilers) control onto a single stick. This device, which uses nylon and teflon through 2024 T3 channel aluminum, insures smooth stick operation in flight. These changes and others have kept the people at Birdman rather busy. But after close to a year spent in re-engineering the TL-1A, a 1000-hour static test was made.

At 122 pounds, the new Birdman is much stronger than its predecessor. It is designed to load factors of plus or minus 6 G's. Birdman TL-1A measures 19.5 feet from nose to its V-nail and 34 feet from wingtip to wingtip. It separates into four parts—wings, center section and tail cone/mepennage—for easy transport.

Birdman TL-1A (Figs.9-3 and 9-4) is sold in kit form complete with the engine for $2,495. A color brochure is available for $5 from Birdman Aircraft Inc., 480 Midway, Daytona Beach Regional Airport, Daytona Beacb, FL 32014.

The Windwagon

Small is beautiful, small is inexpensive and small is sheer fun!

These thoughts occur when you watch Gary Watson's wonderful little 273-pound Windwagon (Fig. 9-5) take off in gusty air, track outbound stable as a far heavier aircraft and return to land like a feather on its rigid tricycle landing gear. The gear is made from simple tubular legs with no shocks other than the tires.

Gary got the idea for Windwagon after converting a VW Beetle engine by sawing it in half, utilizing the best parts and scrounging a different case that was in better shape. It is a real junkyard powerplant that displaces about 900cc and delivers in excess of 30 horsepower.

Gary already had built and flown a Cal Parker Teenie Two. It served as inspiration for much of the aerodynamic design to carry the engine tractor-mounted. The engine has no electrical system. It is fitted with a single Slick magneto and a Posa carburetor and not much more. The gas tank holds four gallons.

In fitting the fuselage to the VW engine, Gary went the easy route in building it small, building it light and building it simple. The fuselage has a circular cross section. The bulkheads were bent up over plywood form blocks with a plastic hammer. All bulkheads fit in upright except the one that serves as the pilot's back rest. That one is slanted rearward to permit a semi-supine seating arrangement. There are no compound curves involved—all the fuselage skins are flat-wrapped.

Windwagon's wings are of constant chord using the Clark Y airfoil, built in three sections. The center section carries the main gear and the two outer panels can be removed quickly and easily for trailering. The spars are built up using strips of flat aluminum sheet for the web and off-the-shelf aluminum angle bars for cap strips. There is no bending required and the only tools needed are a pair of 2 × 4s, a pair of C-clamps, a plastic hammer, drill, bits, tin shears, a

Fig. 9-3. Leonard Roberts took over the Birdman project. He made it very safe with a beefed-up tail and stronger engine mount. This version is called the Birdman TL-1A.

file and pop-rivet equipment. The control system is all pushrod design.

A center mounted control stick is designed to prevent over-control by ham-handed pilots on takeoff. The result is that Windwagon is very stable in ground operation. This was evident when I photographed Gary taxiing down the grass at Oshkosh's Wittman Field during a recent EAA Convention. His technique was good but so was the Windwagon's performance. It behaved like a much bigger aircraft. This was proven on its initial test hop on April 19, 1977, a flight that took place in rough air and came off well.

Fig. 9-4. The new Birdman TL-1A with a spoileron extended for roll control.

Fig. 9-5. Gary Watson's Windwagon is an all metal ultralight with half a VW powerplant.

According to Gary, the Windwagon wants to fly at 40 mph IAS and lifts off easily at 45. Cruise speed is about 90 mph but it will hit 100 mph easily with full bore. Approach speed is 55 mph, IAS power off, and touchdown is around 40 mph. Gary initially used a two-bladed, 50-inch propeller carved by a friend, Dick Bohls, but later went to a 40-inch four-bladed Bohls prop, which provided more ground clearance.

Today Gary sells plans for $50 and a brochure for $5 from Route 1, Newcastle, TX 76372. His advertising reads:

"The amazing 2-cylinder VW Powered Airplane that people are talking about around the world. With the looks of a Jet and more fun than a Cub. Easily built and flown by beginners for pennies."

Not an oversell—Gary spent a bit over $1200 on the prototype. Today, more than a hundred Windwagons are under construction around the world. A big incentive of course is the ready availability of junked Beetle powerplants at low cost. Gary supplies instructions for modifying them for aircraft use.

The Micro-IMP

Molt Taylor, who gave us the Taylor Aerocar, Coot amphibian, IMP and Mini-IMP, has joined the ultralight movement with something new—a "paper" plane he calls Micro-IMP (Fig. 9-6). But let Molt tell it:

The prototype Micro-IMP has been designed to bring builders a modern, efficient, easy-to-build featherweight light plane that can be constructed by anyone reasonably handy with simple shop tools. The Micro is constructed basically from glass-reinforced paper. This new building material greatly simplifies construction since it is easily cut, shaped, drilled, riveted, bonded, sawn, sheared and finished.

The material is readily available and is far less costly than usual aircraft building materials. It is far easier to work with than metal, wood or composites. We plan eventually to have a full kit of all materials and parts to build the Micro-IMP.

A unique feature of the future kit will be that all parts that the builder must fabricate will be printed full-size on the basic construction paper. Therefore, the builder only has to cut the parts out with a sharp knife, using a straight edge. Parts are then suitably glassed using furnished glass fabric and resin. They are then joined mainly by use of simple triangular wood battens and a hand staple gun. The corners are then glassed with glass tape on both sides to make beautiful, workmanlike corners and edges.

The kit will contain all instruments, hardware, engine, shaft system, propeller, landing gear, canopy and the beautiful exterior molded fiberglass skin. All structure is of the basic glass-reinforced paper, including the tail boom, tail surfaces and wings. The wings are easily removable by one person and are so light that they can easily be handled once they are off the fuselage.

Wind covering will be ripstop dacron, similar to that used in many hang gliders, over the structural paper wing frames. The tail surfaces are similarly framed and also covered with lightweight fabric. This type construction is not only much lighter than metals or glass-covered foam, but also is far easier to make and finish. The reinforced paper is suitably protected from ultraviolet radiation in sunlight prior to covering and painting with any colors desired.

Micro-IMP has many unusual and useful features not to be found in other featherweights. An example is a fully retractable landing gear. The retraction is achieved by a simple mechanism of cables operated by a single handle in the cockpit. The landing gear is positively locked in both up and down positions.

Another feature is its GA(PC) NASA airfoil which provides full-span "flaperons" serving both as flaps and ailerons. These surfaces are moved to a reflexed position for high speed cruise with a simple cockpit control. The fully trimmable horizontal tail surfaces permit the pilot to obtain an optimum trim condition for any of the infinitely variable wingflap conditions and give Micro-IMP minimum drag for fantastic cruise performance on minimum horsepower.

Micro-IMP is fitted with the Citroen 2CV engine normally rated at 37 horsepower. However we are limiting the modified engine to 3000 rpm redline and about 25 horsepower for aircraft use. This engine will burn only 1.2 gallons of fuel per hour and

Fig. 9-6. Molt Taylor poses with his Micro-IMP.

Micro-IMP is equipped with a 7-gallon fuel tank built into the structure.

The engine drive-line features the Taylor developed Flexidyne dry fluid coupling. It has been FAA certificated in similar shaft application for the past quarter century, as in the early Aerocar, and eliminates any vibration problems. The Flexidyne has been modified to permit hand-cranking through the shaft system if desired. But a starter can be used along with an alternator.

The modified engine also is equipped with a condenser discharge magneto type ignition system. It uses no points or distributor. The new mags feature electronic spark retard for easy starts and smooth idling. The ignition system is fully shielded for good ratio reception and the solid state magneto gives an exceptionally hot spark for quick starts. In place of common spark plugs, the engine uses surface gap igniters.

The four-stroke engine is extremely light and is completely disassembled, magnafluxed and dye-checked prior to reassembly. The crankshaft is modified for the output flange. The engine is torn down and a special intake manifold and exhaust system is fitted along with an injector-type carburetor featuring automatically compensated adjustment for high-altitude operation.

A two-position propeller has been developed to give Micro-IMP maximum takeoff and climb performance as well as optimum cruise. With an empty weight of only 250 pounds, Micro-IMP will still equal the performance of a Cessna 152. A more detailed report on actual performance will be available following completion of flight tests with the first flying prototype, which should be in the air by the time you read this.

The structural heart of Micro-IMP is the pilot's seat. It is designed so that upholstery and arm rests are simply snapped in. The flight control system of Mini-IMP, with the side controller, rudder-vator mixer and flaperon systems, were further simplified and lightened for Micro-IMP.

Due to its light empty weight compared with its useful load (about 250 pounds), Micro-IMP has been fitted with high aspect ratio wings (27 foot span, 3 foot chord). This, coupled with its exceptionally clean aerodynamic configuration, is responsible for its anticipated excellent flight performance.

We estimate the Micro-IMP Deluxe Kit will cost in the area of $3,000, available in a progressive purchase plan. For further details contact M. B. (Molt) Taylor, Box 1171, Longview, WA 98632.

The Bi-Fly

On the West Coast, a fellow named Robert C. Teman has come up with an interesting ultralight design he calls Bi-Fly (Fig. 9-7). Here is his report:

Bi-Fly took less than a year of weekends and evenings to build, but several years were put into testing various configurations and coming up with solutions to the many problems of a new design. Safety, weight, cost and function had to be considered for each part.

The one-dimensional fuselage structure is 6061-T6 X .090 wall aluminum tubing, 3 inch diameter, bolted together with 2024-T3 gussets. Although the simplest part, it was the most

Fig. 9-7. The Bi-Fly is a new biplane ultralight designed by Robert C. Teman of San Diego.

difficult to design since it fixes the location of all components—the wings, seat, landing gear, engine and tail. This frame must also transmit all primary loads.

All-flying empennage surfaces and four ailerons are conventionally controlled with a yoke mounted on a pedestal and a rudder bar.

Tricycle landing gear with a steerable nose wheel provides good ground handling characteristics and is forgiving in landings. The main gear has a rubber donut type compression suspension and the nose wheel uses a coiled spring. A control pedestal is located between the pilot's knees and provides a mounting for instruments and controls.

An Onan 18-horsepower engine swings a 46 inch diameter by 26 inch pitch propeller at about 3200 rpm. The modifications to the off-the-shelf engine were relatively simple. Thirty pounds were pared from the 100-pound powerplant. The steel cooling shrouds were replaced with a lighter version and the cooling fan was removed along with the electric starter, governor and mufflers. The flywheel was turned down for substantial weight saving. New exhaust pipes were made. After 50 hours running, the engine proved to be economical and reliable. Fuel consumption was running about 1.5 gph.

Taxi testing began on the first of the year 1979. The first flight came as a surprise during a high-speed taxi run at 28 mph. A series of short flights down the 4000-foot runway at Ramona was used to determine flight handling.

The roll axis proved very stable, owing to the high dihedral angle of 6 degrees 30 minutes. This was changed later to 3 degrees in favor of less washout and more lift. Control is responsive but rather slow.

The Bi-Fly will fly straight and track nicely without rudder yaw adjustments. The rudder is sensitive to small movements because of its location in the propeller slipstream. Gentle turns can be made with rudder only by putting the aircraft into a slight yaw and letting the outboard wing pick up automatically, aiding coordination in the turn.

The stabilator provides generous pitch control. But it is also slow to respond because of the low flight speeds, in a 35-mph cruise mode. Takeoff requires about 2 inches of aft yoke and power landings can be flared neatly with about 4 inches aft yoke travel at 25 mph.

The Bi-Fly has at this writing undergone more than 10 hours of logged flight time in preparation for expected FAA certification.

Fig. 9-8. The composite construction Quickie flies on an 18-horsepower Onan engine and gets up to 100 miles per gallon.

Weight-saving improvements and design modifications are, of course, on-going as in any new design. Upon completion of the flight test program, it is anticipated that construction plans will be offered. See Table 9-2. An information package costs $5 and is available from Robert Teman, 10215 Ambassador Ave., San Diego, CA 92126.

Quickie

On August 4, 1978, a strange little biplane—or canard monoplane, depending on your point of view—called *Quickie* won the coveted Outstanding New Design Award from the Experimental Aircraft Association at their annual Oskosh convention. Reasons for

Table 9-2. Bi-Fly Specifications.

Description	Experimental, Single-Place Pusher Biplane.
Engine	Onan 18-hp
Wing	Modified Clark Y
Wing Construction	Fiberglass Expoxy
Span	24′ 3″
Area	124 sq/ft
Chord	31″ (Constant)
Stabilator	All-Flying, 16 sq/ft
Rudder	All-Flying, 10 sq/ft
Empty weight	235 pounds
Gross weight	500 pounds
Fuel Capacity	3.2 Gallons
Fuel Consumption	1.5 gph
Takeoff roll	150 feet
Stall	25 mph
Top Speed	45 mph
Cost to Build	Approx. $1700

the award were given as the pioneering work done on the Onan engine together with an exceptionally efficient aerodynamic design. This combination permitted a low-cost aircraft with top performance for its horsepower.

In flight, Quickie reminds me of a dolphin playing in the sparkling waters of the Sea of Cortez, with agile leaps and splashes. I followed it around the sky one day taking pictures and noted that even its canard wing resembled the flippers of the sea mammal.

Quickie Enterprises of PO Box 786, Mojave, CA 93501, consists of two inspired fellows named Gene Sheehan and Tom Jewett. They decided to build an ultralight aircraft around the rugged little four-stroke, direct-drive Onan engine that operates at a relatively high rpm (3600) continuously. It is used primarily in such applications as recreational vehicle generators. Stripped for flying, the Onan weighs only around 70 pounds and has an overhaul requirement of about 1000 hours.

Jewett got the idea of converting the Onan to aircraft use, because it had more than a million sales to back up its reliability factor. Though a bit heavy for its power output, it became a challenge to design an aircraft to match it. The pair turned to Burt Rutan, designer of such exotic composite aircraft as the VariViggen, VariEze and Defiant. Rutan got busy and worked out a design to offer good performance and construction simplicity with an empty weight of 240 pounds.

Rutan went to the tractor/canard/tailles concept to put the pilot right on the center of gravity and combined the canard and landing gear to offer low drag and less weight. The canard carries a full-span elevator/flap system with inboard ailerons on the rear main wing. The tailwheel fairing substitutes for an aerodynamic rudder.

On November 15, 1977, after 400 manhours work, Burt, Tom and Gene all test flew the prototype Quickie at Mojave Airport. They logged 25 hours on it in the first month. With the design frozen and proven, Rutan went back to work on his Defiant and other projects. He stepped out of the Quickie program to help VariEze builders with their projects.

Flight characteristics of Quickie were good—little adverse yaw, good stall recovery and improved visibility. Hands off flight is stable, even in rough air. Takeoff is unique—there's no rotation. With full aft stick or neutral stick Quickie just levitates. With full forward stick, the tailwheel lifts at 50 mph. But you have to get the tail back down a bit to fly. Fast taxiing is fun with no groundloop

Table 9-3. Quickie Specifications.

Specification	Value
Engine	18 hp Onan
Length	17′ 4″
Wingspan	16′ 8″
Total Wing Area	50 sq/ft
Empty Weight	240 lb.
Gross Weight	480 lb.
Useful Load	240 lb.
Baggage Capacity	30 lb.
Fuel Capacity	8 gal.
Cockpit Length	64″
Cockpit Width	22″
Takeoff Distance (SL)	660 Feet
Landing Distance (SL)	835 Feet
Stall Speed (power off)	53 mph
Stall Speed (power on)	49 mph
Top Speed	127 mph
Cruise Speed	121 mph
mpg at 100 mph	85
Normal Cruise Range	550 sm
Rate of Climb (SL)	425 fpm
Service Ceiling	12,300 Feet

tendency due to the wide gear trend. The mains are at the outer ends of the canard.

Construction is basically of sandwich type using high strength fiberglass and a foam core. No expensive molds or tools are needed. Building time runs about half that of a VariEze. Tom and Gene flew their prototype Quickie back to Oshkosh from Mojave, taking turns over two and a half days, covering 2025 miles in about 19 hours and averaging 65.1 miles per gallon. Total fuel cost was only $30.

A number of other homebuilt Quickies are now flying. Kits are available including engine and electrical system in an incremental program with prices starting around $3000. See Table 9-3. Plans are available for $150, and an information pack for $6. The address is Quickie Aircraft Corporation, PO Box 786, Mojave, CA 93501.

Chapter 10

The Flying Bathtub

In 1931 the nation was still reeling from the shock of the Great Depression. Gasoline was 7 cents a gallon and sirloin steak was 23 cents a pound. Few people had enough loose money to send a letter by airmail. Nevertheless, the country was air-minded and Henry Ford saw a golden opportunity to put an airplane in every garage. An inventor named Harry Brooks induced Henry to pour a bunch of shiny new dimes into a tiny low-winger that promptly was called the Ford Flying Flivver. But it crashed, killing Brooks. Ford went back to making Model Ts.

Ramsey's Tub

The idea caught on, though. A whole bunch of pocket planes appeared briefly—the Alexander Flyabout, the Curtiss Junior, the C-2 Aeronca and an unbelievable flying machine called the Ramsey Flying Bathtub. Plans appeared in the 1931 and 1932 editions of the *Flying and Glider Manual*, granddaddy of all current periodicals devoted to homebuilts.

The Ramsey Bathtub did not make a clean sweep of America's skies, much as we'd like to believe, but a few handy farm boys did get Tubs flying. They had two side-by-side seats with a joystick in between. There were two sets of rudder pedals, but the heel brakes were hooked up only to the outboard pedals—left and right. In the event of an incipient groundloop, say to the left, the pilot would lean over and shout: "Okay, Charley! Push with your right heel!"

Time marched on, the Depression ended and folks got to flying bigger, faster and more expensive airplanes as World War II broke out. The Ramsey Tub was forgotten. In 1969, a couple of college students in Gilroy, CA working for their airframe and power plant licenses, saw a copy of the *Flying and Glider Manual* on the desk of their instructor, Moe Mayfield. The students, Jonathan Teeling and Jose Gonzales, knew right away that would be their dream ship. They got busy building it at the Gavilan College aeronautics department at Hollister Airport. By 1971 it was ready to fly.

They'd made a few changes in the interest of survival and added a Revmaster 1834cc VW conversion for power. Their Tub weighed 550 pounds empty and 910 pounds loaded with Jon and Jose flying. They played around the California countryside for a while, until Teeling got a job as an FAA traffic controller in Panama. The Tub went into Jose's garage in Fairfield. Only recently, with Teeling back home, did they begin to resurrect their toy.

Meanwhile Bob Said, an aviation writer, looked up Teeling and Gonzales to do a story and that caught the eye of a fellow named Ellis Moncrief, in LaGrande, OR. Moncrief contacted Teeling and Gonzales and then wrote to the Experimental Aircraft Association to get a reprint of the 1932 *Flying and Glider Manual*.

He spent a couple of years building most of it, but a job change forced him to put his unfinished Tub in storage. Enter G. Irvin Mahugh, a civil engineer with the U.S. Forest Service. A family man with five children and an aeronautical background at Boeing, Mahugh had worked as test director on the 737 certification program and conducted flight tests of an automatic instrument landing system for the Boeing SST that was never built.

Mahugh was in LaGrange coordinating work on some new fire bombers, when he bumped into Moncrief. "Wanna buy a Ramsey Flying Bathtub?" Ellis asked.

"A what?" Mahugh replied, thinking maybe it was a joke. Moncrief showed him snapshots of the work he'd done—a complete fuselage and empennage all ready for covering, with a basket case Continental A-75—all for $1500. Mahugh checked *Trade-A-Plane* and learned that the engine alone was worth that kind of money.

In February 1977 Mahugh got busy rebuilding the engine, with a new set of piston rings that came by mail. Carefully and thoroughly he rebuilt the engine and uprated it to 85 horses. Then came the big dream—he and his son Jim would fly it all the way from their home at Baker, OR to the big EAA Fly-In at Oshkosh in the summer of 1978!

There was plenty of work remaining—making the metal cowling, engine cooling baffles and shrouds, design and fabrication of the brake system, plus adding a wooden turtleback fairing over the baggage compartment. In April, the first shipment of aircraft spruce and plywood arrived and father and son got busy building a Piper J-3 wing with the same planform as the Ramsey wing. They needed extra lift to carry the beefed-up fuselage and more powerful engine.

Mahugh went to the Clark Y airfoil and built a full-size 63-inch rib jig, with truss braces and spar spacing following the J-3 rib pattern. Truss braces and gussets were kept to a uniform size for ease of building, with each piece numbered and put together like an assembly line production. The gussets were stapled to spruce capstrips, with all wood joints glued with Aero-Lite glue. The staples were removed when the glue dried. All sharp edges were sanded down and the finished ribs coated with polyurethane varnish.

Compression struts consisted of three-fourth inch x .049 inch tubing of 4130 steel with .063 inch plates welded to the ends. Drag wires were of 0.125 inch semi-hard piano wire, looped around cable thimbles at each end. Metal fittings were cut from raw 4130 sheet and tubing stock. Spars were the same size as J-3 spars and made from Sitka spruce. By September 1977, both wings were ready for inspection. Ed Elder, from the Portland FAA GADO, looked them over and signed them off for covering. Press of other matters slowed the project until January, 1978, when Mahugh made the mistake of attaching the leading edge metal on a cold day. This made it change shape later. Now, when he flies the Tub in warm weather, the sheet metal "oil cans" between the ribs and makes the airfoil a bit wavy.

By March, they'd begun to skin the Tub with 2.7 ounce Dacron bonded with Fab-Tac cement, with the help of Bud Bailey. They tried to rush the job but, says Mahugh, the old saying, "The hurrier you go, the behinder you get," held true. Once he dropped the spray gun smack onto a freshly painted aileron and had to do the job over.

Dreams of making the daring flight to Oshkosh faded until Mahugh learned that the FAA had relaxed its policy of requiring 50 hours to be flown in a local flight test area prior to leaving cross-country. The FAA lifted the restriction after only 25 hours, because of the certificated engine.

Mahugh tied the Tub's tail to a neighbor's fence post behind the garage and began the engine runup. By May 8, a pretty day, he was ready to move it to the airport and into a large hangar owned by Jim Hanley, rent free. Mahugh and his wife worked until 2 a.m. getting the plane ready for flight. Finally, Les Briggs, an FAA inspector, stopped by for a look. He found only one thing wrong—an aileron pulley guard installed inside the cockpit needed more clearance. Five minutes of work with a rattail file fixed the problem.

Fig. 10-1. The Ramsey Bathtub was a popular ultralight in 1932. Irvin Mahugh built this replica.

Next came more ground runups and taxi tests that proved the plane handled fine on the ground, despite the close-coupled tail-wheel. The center of gravity was located at 33% instead of 25% mean aerodynamic chord, but a few slow flights off the runway revealed no unusual flight problems.

By May 15, after a rainy spell, the decision was *go*. The Mahughs stopped by school to pick up the children and in-laws arrived from Seattle for the big event. Says Mahugh, "After one last low, slow pass up and down the runway, I decided to go aloft. Liftoff from runway 34 was at 9:27 a.m. for the most exhilerating half hour of my life. What a thrill! Thank God, dreams can come true!"

The flight testing called for a few minor changes. A left wing heaviness was corrected with lift strut adjustment and addition of a rudder trim tab. The wooden 74/78 propeller he'd installed was a poor match for the draggy open framework and only turned to 2100 rpm. Antique buffs were shocked when he switched to a metal prop, 72/38, but it turned up okay and helped shift the center of gravity forward a bit.

On to Oshkosh

By mid-July, Mahugh had the Tub (Fig. 10-2) all cleaned up and ready to fly to Oshkosh with son Jim. At dawn on July 22, they headed for Idaho Falls, with a gas stop at Gooding, ID. Mrs. Mahugh and two daughters raced along behind them in the family car. Despite headwinds, the Tub averaged 72 mph ground speed. They got into Idaho Falls a good two hours ahead of the gals.

From there they headed north, climbing the Tub to 7500 feet to cross the Continental Divide near West Yellowstone. They flew down the Madison River Valley to Bozeman, MT where Mahugh had gone to college. Clear skies and a brisk tailwind the next morning helped the Tub along as Mahugh and son wound through the Big Sky country. Following mountain passes, they made stops at Roundup, Glasgow and Fort Peck, where the pilot had lived as a boy and where he'd first soloed a PA-11, 28 years before.

Eastward across the flat Dakotas and into Minnesota's lake country they flew, wide-eyed at the beauty of America as seen over the rim of their Bathtub. At Olivia, a friendly mechanic helped fix a broken cowl clip over the number four cylinder and wouldn't charge a cent. They finally rendezvoused again with the gals at Winona and spent the night at a KOA campground. The next day they flew off for the final lap to Oshkosh.

The Fly-In was a huge success, the more so when their Tub won a big prize—the Vintage Aircraft Outstanding Replica Award. It was there I met Mahugh and son and took some photos. When the day ended Mahugh headed west. The return flight was routine with only one weather hold. Mahugh flew the Tub back alone, leaving Jim to visit relatives in Ohio with the gals.

The Wier Draggin' Fly

There I was at 1000 feet over the green countryside, rain spattering my face in a cold shower while I aviated around the sky in a crazy machine that is fondly called a flying bathtub. All I needed was a cake of soap and the lung power to sing lyrics into the wind like; "Off I go, into the wild blue yonder!"

Courtesy of Ronald D. Wier, former president of EAA Chapter 14 in San Diego, I was having myself a ball flying the nearest thing to Amos N' Andy's old Fresh-Air Taxi Cab. Officially it's called the RDW-2, Serial No. 1, and otherwise known as the Wier Draggin' Fly (Fig. 10-2). But it's much much more!

Aerial Simplicity

Now with a new owner, near Los Angeles, the Draggin' Fly is aerial simplicity itself—one of the first of the Microlights. It has a wing to hold you up, a VW engine to make it go, a tail to steer with, a funny tub-like place to sit, one magneto switch, one go-lever, a push-pull carb heat control and a stick and rudder. Beyond that, nothing much except a tricycle gear of three go-kart wheels to taxi around on and a few rudimentary gauges to tell how high is up, how fast (or slow) you're moving, whether you're slipping or skidding, and whether the engine room is functioning. Trim tab? Who needs it? Brakes? What for? Radio? Why talk when you can sing?

Fig. 10-2. Ron Wier's Draggin' Fly has tricycle landing gear, a tractor powerplant and it can fly at 60 miles per hour.

I did a double take the first time I saw the Wier Draggin' Fly. I became completely fascinated when I watched Ron Martin, a local certified flight instructor, driving down Ramona Airport with a big grin and go screaming off clawing for altitude. Another happy chap standing nearby with a big EAA patch on his jacket wandered over, and introduced himself as Ron Wier.

Ron Wier is the kind of guy who belongs singing in a bathtub, pardon the metaphor. A tub with wings yet, a fun loving guy who simply did what most of us would like to do but never get around to—giving wings to our imagination with fabric, wire, tubing and a run-out auto engine.

I asked Wier how long it would take to draw up the plans. I was mentally figuring how I'd hook up a Honda to an ironing board and go sailing through the nearest TCA screaming *beep! beep! beep!* to clear the area of big stinky jetliners.

"Hah," Ron replied. "What plans? There weren't any. Only drawing I did was for the rib jig. Eyeballed the whole darned thing!"

When I raised an eyebrow, my left one, Ron admitted he'd been just a little bit influenced by that funny old flying machine from back in the Roaring Twenties—the Ramsey Flying Bathtub.

"I kept seeing these new Volkswagen conversions everywhere," he went on. "I'd done some work for Ladisalao Pazmany and Bud Evans, on their PL-4A and VP-2 machines, but I wanted something real different. So I dug out an old manual and flipped through the pages until my eye caught the Ramsey design. That was it—why not modernize it?"

The Ramsey Flying Bathtub might not be as well known around airports as the Cessna 152 or Piper Cherokee, but once you've seen one you'll never forget it. It has a boat-shaped single-holer with an

Evinrude outboard motor up front and a converted OX-5 water pump hooked up to a water radiator.

The Draggin' Fly, of course, doesn't need a water pump or a radiator since the VW engine is air-cooled. At first Ron stuck in a tiny 36-horsepower VW engine. But after eight hours of trying to get off the ground at a gross weight of 450 pounds on a hot day, he went to the bigger 1600cc engine.

That turned the trick. The gross rose to 680 pounds with the replacement engine and some other modifications plus a full tank of gas and a 170-pound driver. She really flew! Ron left the newer engine as stock as possible but added a Ted Barker propeller hub and carved his own club out of a chunk of mahogany in just three days. Being semi-retired after 19 years as a Dunn & Bradstreet stock analyst and two years as a Lt. j.g. aboard a Navy carrier, he had the time and the background for the big challenge—getting a bathtub to fly.

"I used to hangar fly a lot with Navy jocks and I learned a lot of aerodynamics that way," he recalled. His self-education including much reading of books on airplane design. Finally, he got busy and built himself a Stits Skycoupe. Ron started flying at 14, bootlegging some J-2 time and went on flying on a private ticket, to hell with the ratings. He considers flying a fun pasttime, not a chore.

What he really had in mind, though, was something simple to fly, easy to land and drive on the ground, and ultra-safe. With the Ramsey Tub in the back of his mind, he looked up the shape of the Piper Cub wing—a USA 35-B airfoil—which was renowned for its gentle stall and excellent low-speed characteristics. From there he was on his own.

Starting at the bottom, he decided on a trike gear for easy ground driving, like a baby buggy. Up front, he made the nose wheel steerable and shock mounted. He bought three go-kart wheels—10 and one-half inch diameter 3.50 by 4.00s with no brakes. He did install a friction parking brake on the front wheel only.

After shaping the bathtub part to fit him, Ron began welding up a bunch of chromoly tubing that stretched back to hold up the tail. Both aileron and rudder action was linked by push-pull rods, with cables running back to the rudder. The open framework left most of the control system exposed, excellent for pre-flight inspection—utter simplicity winning out. The fuselage part was assembled on the floor of the Wier garage. Ron simply drew himself a chalk line on the floor and filled it in with one-inch tubing left over from an old Waldo Waterman Wright Flyer project.

The fuselage, wing and tail surfaces were all fabric covered with 2.7-ounce Dacron glider cloth. Wing spar was made from three-fourth inch spruce plank, the rib cap strips from one-fourth inch square spruce and the ribbing of one-fourth inch ply. The engine installation is such that you don't have to wrestle the cowling off to change the oil—there isn't any cowling.

To fancy things up a bit, Ron stuck on a curved windshield of unbreakable Lexam, a DuPont product. Beyond that, you're on your own in fresh air sitting at midchord under the parasol wing. Looking left and right, you can see the ailerons stretch out full span. When you open the throttle to go, you feel the flippers go solid and effective with gentle pressures on the short stick.

The ease of control is the surprising thing about flying Draggin' Fly—a responsive roll rate, good elevator action and plenty of rudder. There's no buffet and the stall is Cub-like. It has a break straight ahead with no roll-off tendency.

"Fly her wide open all the way!" Ron yelled over the bark of the engine when I taxied out to the runway. I nodded. The tach wiggled around 3300 rpm and stayed there. The airspeed needle hung between 60 and 70 mph on climb, cruise and letdown. At Ron's further suggestion, I left on about half carb heat due to rain and low temperatures that were the order of the day and I used a power approach.

I did try a couple of power-off stalls at altitude, got a clean break at 35 mph indicated airspeed and I felt there'd be no problem on landing. There wasn't—I simply flew it onto the deck at 60 mph, chopped power and slowed to a stop, all in maybe 200 feet. That convinced me—here was the nearest thing to riding a motorbike with wings. A compact little job with a span of 24 feet 5 inches with a 54-inch chord (including 6 inches of aileron), 110 square feet of wing area, a length of 17 feet 5 inches, height to tip of rudder 6 feet 8 inches and a gross weight of 680 pounds.

Though Ron's Draggin' Fly has since been sold to a couple of happy chaps in the Los Angeles area, Ron still sells sets of plans for $20 per set prepaid. Interested? Drop him a line at: Ron Wier, 6406 Burgundy, San Diego, CA 92120.

Chapter 11
STOL Aircraft

"Charley, would you please fly up to the roof and check the shingles? I think I heard a reindeer walking around up there last night!"

Far-fetched? Maybe—but you never know what the future holds for some of the wild contraptions builders are coming up with in the ultralight field. Like a new "hang helicopter" recently developed by a veteran helicopter designer, Webb Scheutzow, of Berea, OH. It is a device he calls a "treetop" one-man helicopter that is foot-launched. Scheutzow says that it "perhaps opens entirely new vistas in sport aviation, as well as in practical applications."

The Stork

Stork is the appropriate name given the long-legged ultralight (Fig. 11-1) Scheutzow developed after some 25 years as an active member of the American Helicopter Society. He also developed the FAA Certificated Scheutzow Model B Helicopter and a number of other exciting whirlybirds.

"Stork is controlled by weight-shift," he explains. "You might call it a 'hang-helicopter.' A first of a kind. Powered by a snowmobile engine, it employs a type of main rotor with special gyro stability qualities. It is foot-launched, like a hang glider, but into a hover attitude."

Scheutzow points out, "although many attempts to build a successful back-pack helicopter have foundered, I made a technical

Fig. 11-1. Webb Scheutzow's Stork is an ultralight, foot-launched helicopter designed in 1979.

study of the requirements for this type of helicopter. My conclusions were that if we are to have a successful helicopter of 70 or 80 pounds, the engine must weigh not more than 12 to 15 pounds and have a rating of 20 horsepower. And there is no readily available engine of this kind."

Nearest thing to this requirement, he says, is the Herbranson RPV engine, which is expensive and not readily available for helicopter use. Expanding his study further, Scheutzow learned that during the 1950s several successful single-place helicopters were built in the 30 to 40 horsepower bracket, with empty weights in the range of 275 to 400 pounds. These light choppers had been built in a Marine Corps competition and were referred to as "rotorcycles." Three examples are the Hiller XROE-1, the Goodyear Gizmo and the Del Mar Whirly Mite.

The Scheutzow Stork development fits between the rotorcycle and the back-pack helicopter categories. It is similar in size and power to the rotorcycle, but has a considerable lower empty weight and foot-launch capability compared to the back-pack concept.

Stork is designed to carry a maximum useful load of 250 pounds, but is not power-limited and could prove capable of lifting a heavier load, the designer says. Final gross weights will be determined on the basis of safe handling characteristics proven in a rigorous flight test program that had not yet been completed at this writing.

Design considerations for the Stork include:

- Weight-shift control.
- Ability to hover by partially loading the rotor and, like a hang glider, learning its handling characteristics by "ground flying." Long fiberglass skids, which are removable, function like bicycle "learning wheels."
- Important weight savings derive from its being foot-launched and landed.
- The FAA does not require formal licensing of foot-launched aircraft or their pilots at this time. However, that could change in the future.
- During the past five years, thousands of people have learned to use control hang gliders by shifting weight.

Stork's main rotor has two blades mounted "rigid-in-plane." The blades are mounted on offset flapping hinges with a "delta-three" angle. This provides automatic pitch control for both power-on and power-off autorotation flight and also provides a "flat-tracking" rotor. There is very little change in rotor attitude during gust conditions, says Scheutzow.

The dynamics of Stork's configuration provide for stable, long-period motions suitable for weight-shift control. The control bar has twist-grip throttle control for the left hand and tail-rotor pitch control for the right hand. An arrangement is made so that the two can be synchronized or controlled separately. Patents have been filed covering all the Stork's new features.

"The Stork," says Scheutzow, "is truly a low-cost helicopter—something that many have attempted previously, but which no one, including myself, has until now delivered. Potentially, a quality assembly kit for Stork can be produced and market at motorcycle prices."

Webb Scheutzow's helicopter career goes back to 1944 with Kellett Aircraft in Philadelphia. There he contributed to the engineering design of the first twin-engine transport helicopter—the XR-1. Subsequently he was employed at Hiller Helicopters in Palo Alto, CA where he participated in design and engineering of the original overhead stick "A" Model H-23. After 11 years as a test and development engineer with General Motors' Cadillac Division, he formed his own company and developed the Scheutzow Model B utility helicopter.

The Model B got its FAA Type Certificate in 1976. In 1977, the Scheutzow Helicopter Corporation was sold and moved to

Texas. Since then Scheutzow has turned to design and development of ultralight aircraft. An earlier project, the homebuilt Hawk 90 and Hawk 140 helicopter program, also has been shelved. Stork, his latest project, looks like a real winner because of all the current interest in ultralight aircraft. An information kit on Stork is available for $6 U.S. ($7 foreign) from Webb Scheutzow, 451 Lynn Drive, Berea, OH 44017.

The Beta Bird

Some guys just aren't happy to let well enough alone. They come up with a great idea for a flying machine, build it, fly it and sell thousands of sets of plans to happy homebuilders the world over. Then suddenly they make a phone call, like one I recently got: "Don, hurry up to Mojave Airport this weekend! You gotta see my new bird fly!"

I recognized the voice as that of Bob Hovey, the aerospace engineer who took time off back in 1970 to design the delightful little Whing Ding, maybe the world's smallest ultralight biplane (Fig. 11-2).

It is a fun little job that weighs only 123 pounds before you fill it up with gas. This open-air pusher can go 50 miles per hour on a calm day.

I wrote a magazine report on Whing Ding and the response was overwhelming. Bob said he sold some six thousand sets of plans at $20 a set. He also used more of his spare time to write several books, on how to make propellers, how to design a ducted fan and how to run stress analyses on all sorts of ultralights.

It was no surprise to hear that Hovey had done it again and come up with a new design he called the *Beta Bird*. It weighs 405

Fig. 11-2. Bob Hovey's Whing Ding biplane is popular. More than 6000 sets of plans have been sold.

pounds dry, a bit heavier than Whing Ding, and it only has one wing. The engine is a converted 1385cc VW that puts out around 45 horsepower, swinging a 54-inch diameter prop with a 24-inch pitch.

"Carved it myself!" Bob said proudly. "Followed directions right outa my propeller book!"

So what's a Beta Bird?

The name seems a bit premature. Or at least incongruous. It refers to a special propeller that Hovey was still working on, which was not included in the initial set of plans. It would be big, with maybe an 80-inch diameter, and controllable—though not constant-speed. The idea is to maximize Beta Bird's low-speed performance to give it an amazing versatility as a short takeoff and landing aircraft that you can operate off a dime, or at least a quarter.

The idea for Beta Bird (Fig. 11-3) came to him after learning that Whing Ding had proven highly popular, not as a toy but as a practical flying workhorse he should have called Pegasus. Farmers loved to use it for checking the south forty, or rounding up cattle. In Australia, the outbackers were really turned on by the idea of having a small, inexpensive, easy-to-build and easy-to-fly aircraft they could fly low and slow while counting koalas in the acacia wattles, chasing kangaroos or whatever they do down under.

What they really needed, Hovey decided, was a more practical plane that could fly better than a mile a minute, handle more easily and have all that good short takeoff and landing (STOL) stuff. To achieve the latter, he decided on full-span "drooperons"—a word he coined to explain their dual function as ailerons and flaps. Hovey designed a neat mixing setup where the control stick wiggles the ailerons differentially and a manual lever on the left side of the seat operates the full-span surfaces together as droopy flaps.

The drooperons have a wide chord of 14.5 inches, or 34.6 percent of the wing's 3.5-foot chord. In the full up position they are nicely faired. They're built of aluminum tubing and fabric covered of 13 percent thickness. Beta Bird's central pylon structure and empenage are both of simple aluminum sheet construction that is pop-riveted to aluminum tubing. The two are connected by a two-inch aluminum boom.

Although the pusher engine (a McCulloch 101) of the Whing Ding is mounted high with the thrust line behind the trailing edge of the upper wing, Hovey mounted the VW powerplant on Beta Bird below the single wing. The thrust line is roughly behind the pilot's head position. This blows the slipstream nicely back over the vertical tail's 8.3 square feet of surface. Horizontal tail surface area is about twice as large—17.8 square feet. Elevator travel is 25 degrees up and down.

The pilot sits up front with all the world to look at through rose colored glasses or a windscreen. He found the most comfortable cruise speed to be an easy 60 mph, although it will do 70 mph wide open with the VW shaft and prop both turning at 3800 rpm.

When Hovey installed the windscreen he ran some unusual "tuft" tests by attaching a feather to the end of a long stick and holding it forward of the craft as he flew, moving it from side to side. He found that the airflow separated around the windscreen in a sort of laminar fashion and then came back together behind the pilot.

When Beta Bird's beta prop is installed, it will be controllable with a lever on the instrument panel and have a choice of several pitch selections—according to Hovey's thinking. These will run from flat pitch to takeoff, to cruise and reverse on flareout in order to permit a zero-speed landing.

On takeoff, the beta prop should wind up from 4500 rpm in flat pitch and zero thrust, with the pilot smoothly adding pitch to the blades as required for an accelerated launch. The prop will be geared to the engine with a belt drive running off a jackshaft. This way the propeller can turn more slowly and keep the tip velocity subsonic.

A Simple Ultralight

As part of the ultra-simplicity of design and construction, Beta Bird uses stock wheels and brakes from a Cessna 150. No tailwheel springs are required. Rudder pedals are mounted to the left and right sides of the front end. A faired body houses the instrument panel with all the dials you'd ever want for a nice VFR flight to nowhere in particular. There's a magnetic compass, sensitive altimeter, airspeed indicator, tachometer, oil temperature and pressure gauges and cylinder head temperature. Throttle is at the left side.

The cylinder head temperature is a must, Hovey feels, to insure that the engine doesn't overheat when flying low and slow on a hot day. He initially did have a heating problem, but he fixed it with a modified oil cooler and aluminum baffles wrapped around the jugs.

So there I was, at Mojave Airport on a pretty summer day, watching Hovey flight testing his Beta Bird at the same locale

Fig. 11-3. Bob Hovey's Beta Bird is built largely of aluminum and styrofoam. It uses a VW pusher engine for power.

where he'd checked out the Whing Ding seven years before. This time he didn't carry his airplane out to the ramp over his shoulder. He drove it out—first-class. There was a mighty-mouse roar from the VW engine. He was off and flying the pattern low enough for me to shoot pictures and prove it was for real and not just some dream machine.

I had the thought that here is a real ultralight airplane. Something you'd feel comfortable flying that was sensitive enough on the controls to behave the way a real airplane should, not just a Rogallo hang glider with a tiny lawn mower engine stuck on behind. A genuine, first-class little machine that seemed destined to outclass the little Whing Ding.

With a design gross weight of 630 pounds, Beta Bird seems destined also to fly its way into the hearts of a whole new bunch of builders, not as a toy but as a practical little plane for ranchers, farmers and just plain outdoorsmen who love to explore the back country from on high. With its short takeoff and landing capability, it seems to be a go-anywhere machine. Hovey says it'll take floats to add water-flying versatility to its capabilities.

The airfoil is Hovey's own design and I can assume he got out his book, *Ultralight Design*, to plot the curve. The design is a variation of the venerable old Clark Y, of Virginius Clark, modified to take drooperons. There is a slotted flap arrangement to provide good low speed control down to flight level zero.

And why is it called Beta Bird? Beta, Hovey reminded, is not only the second letter of the Greek alphabet and a member of the goosefoot plant family, it also is an engineering term for blade angle that eggheads use when they get to yakking about propeller design. And that's really what it's going to be all about later on.

I remember well the day Hovey tested his Whing Ding a good three feet off the deck and later reported: "There was this little pitch instability. I experienced some buffeting over the horizontal tail on takeoff, which led to momentary pitch hunting."

"What did you do?" I asked eagerly.

"What any test pilot would do," he replied calmly. "I analyzed the situation, considered my options and did not bail out. I leaned forward."

In such ways are new concepts, like Beta Bird, turned into reality instead of remaining a drawing board wonder. To make it all even simpler, Hovey designed Beta Bird so that the wings can be folded back over the tail surfaces by one man in order to road-tow it home.

Plans for Beta Bird are available for $60 a set from: Aircraft Specialties Co., Box 1074, Canyon Country, CA 91351.

Index